ELECTROTECHNOLOGY VOLUME 5

LOW-TEMPERATURE PLASMA TECHNOLOGY APPLICATIONS

ELECTROTECHNOLOGY VOLUME 5

LOW-TEMPERATURE PLASMA TECHNOLOGY APPLICATIONS

ROBERT P. OUELLETTE
MARCEL M. BARBIER
METREK Division of the MITRE Corporation
Mclean, Virginia

PAUL N. CHEREMISINOFF
New Jersey Institute of Technology
Newark, New Jersey

ANN ARBOR SCIENCE
PUBLISHERS INC / THE BUTTERWORTH GROUP
P.O. BOX 1425 ▪ ANN ARBOR, MICHIGAN 48106

230 Collingwood, P.O. Box 1425, Ann Arbor, Michigan 48106

Library of Congress Catalog Number 80-65514
ISBN 0-250-40375-7

Manufactured in the United States of America

FOREWORD

This volume continues the series of ongoing studies surveying the status of electric technologies, initiated and sponsored by Electricité de France and prepared by the Metrek Division of MITRE Corp., the objective of which is to see rapid introduction of technology using electricity and to diffuse information favorable to its development.

This study presents the fundamentals of the corona phenomena and reviews the spectrum of applications. Processes adding value to surfaces by low-temperature plasma discharges are described for polymeric surfaces and polymer deposition. Organic synthesis through the effect of singlet molecular oxygen and other plasma applications in chemical organic synthesis are examined.

We acknowledge Electricité de France for its encouragement in diffusing this information to the technical community. In particular, we are grateful to Mr. J. Bouchet, Direction Générale, Service d'Etude et de Promotion de l'Action Commerciale, Electricité de France.

Robert P. Ouellette
Marcel M. Barbier
Paul N. Cheremisinoff

Robert P. Ouellette is Technical Director of the Environment division of the MITRE Corporation. Dr. Ouellette has been with MITRE since 1969 and had been Associate Technical Director since 1974. Earlier, he was with TRW Systems, Hazelton Labs., Inc. and Massachusetts General Hospital. He graduated from the University of Montreal and received his PhD from the University of Ottawa. A member of the American Statistical Association, Biometrics Society, Atomic Industrial Forum and the NSF Technical Advisory Panel on Hazardous Substances, Dr. Ouellette has published numerous technical papers and books on energy and the environment. He edited the comprehensive surveys, *Electrotechnology*, Volumes 1 and 2, published in 1978 by Ann Arbor Science Publishers, Inc.

Marcel M. Barbier is a staff member in Environmental and Safety Systems at Metrek, a division of the MITRE Corporation. Before joining MITRE in 1975, Dr. Barbier was with Science Applications, Inc., served as Research Physicist at the European Organization for Nuclear Research, Geneva, Switzerland, and was Development Engineer at the CSF Company in Paris, France. Dr. Barbier is a graduate of the Swiss Federal Institute of Technology in Zurich, and of the University of Paris, France.

Paul N. Cheremisinoff is Associate Professor of Environmental Engineering at New Jersey Institute of Technology. He is a consulting engineer, and has consulted on environmental/energy/resources projects for MITRE Corporation. A recognized authority on pollution control and alternative energy technologies, he is author/editor of many publications, including the following Ann Arbor Science books on pollution and energy: *Pollution Engineering Practice Handbook; Carbon Adsorption Handbook; Environmental Impact Data Book; Industrial & Hazardous Wastes Impoundment; Environmental Assessment & Impact Statement Handbook.* Mr. Cheremisinoff is also Engineering Editor of *Water & Sewage Works,* an international journal.

CONTENTS

TABLES

FIGURES

CHAPTER 1

INTRODUCTION–FUNDAMENTALS ON THE CORONA PHENOMENON AND SPECTRUM OF APPLICATIONS

INTRODUCTION

Although plasma phenomena have been observed since antiquity, scientific inquiries were not made for many centuries, and industrial applications are rather recent. The premodern history of corona includes Benjamin Franklin's experiments in electrostatics around 1740, Faraday's discovery of electromagnetic induction in 1831, and the construction of the Wirmhurst generator in 1878.

After 1900, rapid progress was made in numerous industrial applications. Two key dates are 1906, when Frederick G. Cottrell invented the first electrostatic precipitator device, and 1935, when Chester Carlson started his search for a better copier. Table 1-1 identifies the key dates from antiquity to the present in the field of corona effect.

Table 1-1. History of Corona Effect

Date	Scientist	Finding
1600	William Gilbert	Published "De Magnete"
1740	Benjamin Franklin	Electrostatics
1831	Faraday	Electromagnetic induction
1906	Frederik C. Gottrell	Electrostatic precipitator (ESP)
1923	Detroit Edison Co.	First installed ESP
1935	Chester Carlson	Electrostatic copier
1944	Battelle Memorial Institute	Development of process eventually became Xerox

CORONA IN CONTEXT

Corona is basically an electrostatic phenomenon. The corona discharge can also be said to be a plasma state.

Electrostatics is defined as ". . .that class of phenomena which is recognized by the presence of electrical charges, either stationary or moving, and the interaction of these charges, this interaction being solely by reason of the charges."

Plasmas are usually defined in terms of kinetic energy or temperature (°K) and particle density or pressure (particles/cm^3). Table 1-2 characterizes a number of natural and manmade plasmas in terms of the two key parameters. The corona phenomenon (labeled glow discharges in the table) is a relatively low-temperature and low-pressure phenomenon. These same plasmas are shown in their density-energy relationship in Figure 1-1. Figure 1-2 classifies processes of production of ions and electrons according to their cause (Nasser, 1971).

Table 1-2. Ranges of Particle Density and Temperature for Various Types of Plasmas (Sheer, 1968)

	Particle Density (No. /cm^3)	Temperature (°K)
Natural Plasmas		
Stellar interiors	10^{22}-10^{25}	~10^8
Stellar atmospheres	10^{10}-10^{16}	10^4-10^6
Nebulae	10^3	10^4
Interstellar space	1-100	10^2
Earth's ionosphere	10^{10}-10^{12}	10^2-10^3
Manmade Plasmas		
Thermonuclear plasma	10^{12}-10^{14}	10^8-10^9
Constricted arc plasma jets	10^{16}-10^{18}	1-5 x 10^4
Free-burning electric arcs	10^{16}-10^{17}	7-10 x 10^3
Combustion flames	10^{16}-10^{18}	3-5 x 10^3
Low-pressure arcs	10^{10}-10^{12}	1-3 x 10^3
Glow discharges	10^{10}-10^{12}	300-600

Atoms or molecules can absorb energy in sufficient quantity to lose an electron. The atoms or molecules become positively charged and are said to be ionized. The product of ionization is one electron and a positively charged ion. The positive ion may lose one or more electrons and become multiply ionized. The liberation of an electron from a gas is termed ionization. The liberation of an electron from a solid is called electron emission.

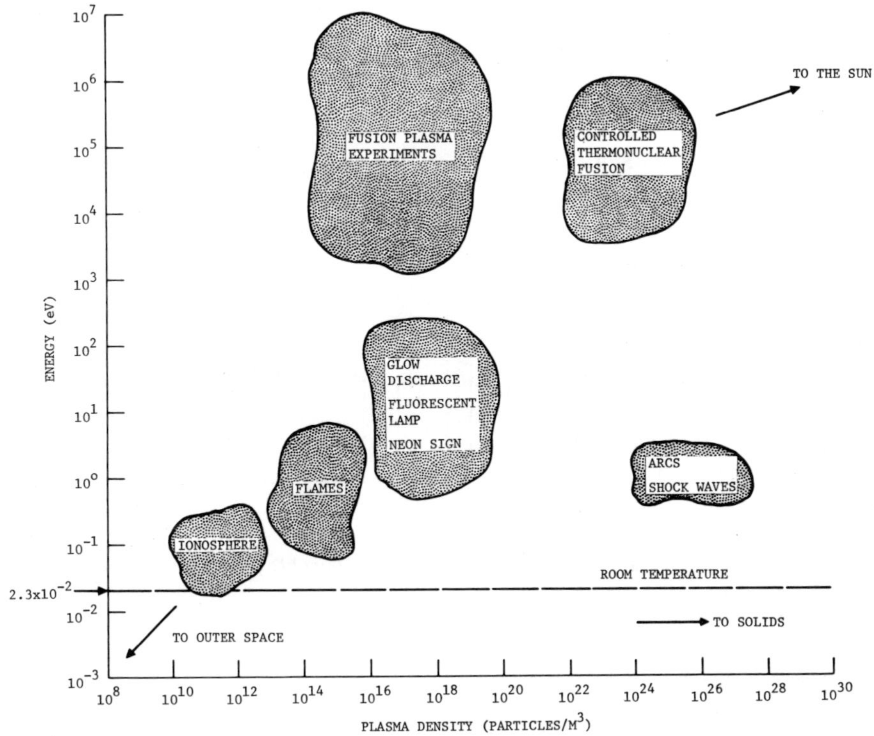

Figure 1-1. Energy-density diagram of the plasma state (Shuhet, 1971).

In gaseous plasma, the ionization of gas by electron collision and by absorption of energy (photoionization) are the most important ionization processes. The electron emission from solids, especially electrodes, contributes greatly to the number of charged particles in a plasma. Also, when free electrons attach themselves to neutral atoms or molecules, negative ions are formed.

A plasma is technically defined as a neutral body of matter containing an appreciable concentration of mobile charge carriers (electrons and ions) in which the Debye length is smaller than any boundary surrounding the plasma.

Plasmas are classified according to pressure, charged-particle density, temperature, boundary conditions, and the presence of external electric and/or magnetic fields.

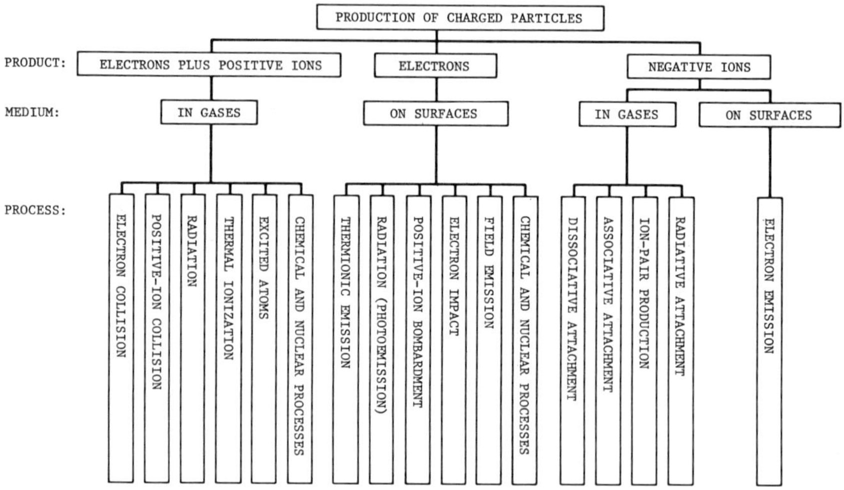

Figure 1-2. A schematic diagram showing the major possible processes for producing the various kinds of charged particles (Nasser, 1971).

FUNDAMENTAL PRINCIPLES

In a discharge, free electrons acquire energy from the imposed electric field and lose it through collisions with neutral gas molecules. This transfer of energy leads to the formation of a variety of species: metastables, atoms, free radicals and ions.

As previously stated, a plasma must be neutral. This occurs when the dimensions of the discharge volume (Λ) are significantly larger than the Debye length (λ_D):

$$\lambda_D = \left(\frac{Eo\,K\,T_e}{n\,e^2}\right)^{1/2}$$

where o = permitivity of free space
K = Boltzmann constant
T_e = electron temperature
n = electron density
e = charge of the electron

Many natural and manmade plasmas are possible, as shown on the energy density plot (Figure 1-3).

In this book we are basically interested in plasmas produced in "glow discharges," that is, plasma characterized by electron energies in the range of 1-10 eV and electron densities in the range of 10^9-10^{12} cm^{-3}. Such plasmas are further characterized by a lack of thermal equilibrium; T_e/T_g in the range of 10-10^2 between the electron temperature (T_e) and the gas temperature (T_g). Actually,

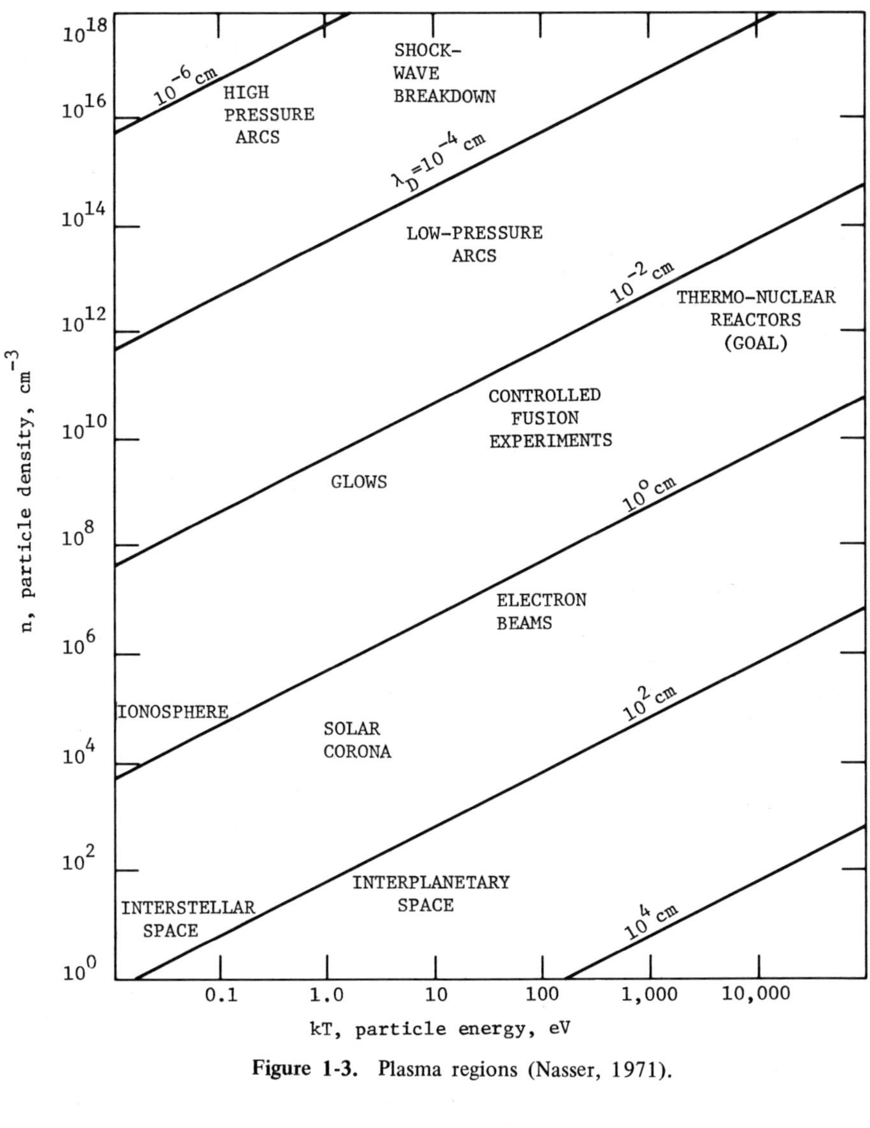

Figure 1-3. Plasma regions (Nasser, 1971).

these plasma can have a gas temperature near ambient while retaining an electron temperature sufficient to rupture molecular bonds.

It is essential to know the electron velocity distribution to derive the electron-energy distribution, the average electron density, the electron-transport properties, and the rate constant of reactions involving electron-molecule collisions.

The shape of the electron-velocity distribution function depends on the magnitude of the applied field and the nature of the elastic and inelastic collisions involved.

Defining f as the density of the electron function, $\bar{r}$ as a space vector, and $\bar{v}$ as a velocity vector, the function may be expressed as $f(\bar{r},\bar{v})$, then $fd\bar{r}dv$ denotes the number of electrons located in the volume element $d\bar{r}$ and whose velocities lie within the volume element $d\bar{v}$, where:

$$d\bar{r} = dxdydz$$

$$d\bar{v} = dxdydz,$$

then $fd\bar{v}$ is the number of electrons per unit volume $\bar{r}$ with velocities $d\bar{v}$, and the electron density at the point $\bar{r}$ is given by

$$n = \int_{-\infty}^{+\infty} fdv$$

where the velocity $\bar{v}$ can be expressed as

$$\bar{v} = V + \langle\bar{v}\rangle$$

velocity	=	random velocity	+	translational velocity of electron clouds

we can define

$$\langle v_x \rangle = \frac{1}{n} \int_{-\infty}^{+\infty} v_x \, fd\bar{v}$$

Since the average value of $\bar{V}$ is zero, the total translational kinetic energy can be expressed as

$$\xi = \frac{1}{2} m\langle v^2\rangle = \frac{1}{n} \int_{-\infty}^{+\infty} \frac{1}{2} m\bar{v}.\bar{v}f.d\bar{v}$$

$$= \frac{1}{2} m\langle V^2\rangle + \frac{1}{2} m\langle v^2\rangle$$

= kinetic energy of randon motion of electrons + kinetic energy of translation of electron clouds

where v and V are the amplitudes of $\bar{v}$ and $\bar{V}$.

Where the velocity distribution is maxwellian, the random kinetic energy can be related to the electron temperature by

$$\frac{1}{2} m\langle v^2\rangle = \frac{3}{2} kT_e$$

The equation summarizing the net rate of transfer of electrons from the volume is the Boltzmann equation expressed as

$$\frac{\partial f}{\partial t} + v \cdot \nabla_r f + \frac{eE}{m} \cdot \nabla_v f = \left(\frac{\partial f}{\partial t}\right) \text{coll}$$

Local variation of the distribution function with time + variation in distribution function resulting from electrons moving in and out of the volume + variation of the distribution function as a result of the applied field

To obtain the velocity distribution function, the equation must be solved. Approximate solutions are available for a number of separate cases. The electron velocity distribution can then be used to evaluate the transport properties of the electrons.

A broad spectrum of chemical reactions occur in plasma. The nature of the gas in the gap strongly affects the number of quantities of species formed. Because oxygen is present in ambient air and because it is a gas often used in plasma chemistry, we give in Table 1-3 some of the most important chemical reactions found in oxygen discharges with their reaction rate constant and with the momentum cross-section probability of elastic collisions.

The reaction rate constant can be defined as

$$K_i = \int_o^\infty \left(\frac{\xi}{2m}\right)^{1/2} \sigma_i(\xi)\ f(\xi)\ d\xi$$

where ξ = electron energy
σ_i = cross section for i-th process
$f(\xi)$ = electron energy distribution function

The effective ion temperature (T_i) as a function of the applied field can be illustrated as follows:

$$\frac{3}{2}\ kT_i = \frac{1}{2}\ (M_i + M)\ v_d^2 + \frac{3}{2} + kT_g$$

where T_i = the ion temperature
M_i = mass of the ion
M = mass of the neutral gas molecule
v_d = drift velocity of the ion

v_d is a function of E/p.

The value of E/p in a plasma plays an important role in determining plasma properties and chemical reaction rates. Hence, it is important to predict E/p under steady-state and breakdown conditions.

Breakdown of a gas in an electric field occurs when the rate of ionization balances the loss of electrons by various processes.

Breakdown will occur when

$$v_i = D_e/\Lambda^2$$

Table 1-3. Elementary Reactions Occurring in an Oxygen Discharge[a] (Hollahan and Bell, 1974)

Reaction	k	$\sigma_{max}(cm^2)$
Ionization		
1. $e + O_2 \rightarrow O_2^+ + 2e$		2.72×10^{-16}
2. $e + O \rightarrow O^+ + 2e$		1.54×10^{-18}
Dissociative Ionization		
3. $e + O_2 \rightarrow O^+ + O$		1.0×10^{-16}
Dissociative Attachment		
4. $e + O_2 \rightarrow O^- + O$		1.41×10^{-18}
5. $e + O_2 \rightarrow O^- + O^+ + e$		4.85×10^{-19}
Dissociation		
6. $e + O_2 \rightarrow 2O + e$		2.25×10^{-18}
Metastable Formation		
7. $e + O_2 \rightarrow O_2(^1\Delta_g) + e$		3.0×10^{-20}
Charge Transfer		
8. $O^+ + O_2 \rightarrow O_2^+ + O$	$2 \times 10^{-11}\ cm^3/sec$	
9. $O_2^+ + O \rightarrow O^+ + O_2$		8×10^{-16}
10. $O_2^+ + O_2 \rightarrow O_3^+ + O$		1×10^{-16}
11. $O_2^+ + 2O_2 \rightarrow O_4^+ + O_2$	$2.8 \times 10^{-30}\ cm^6/sec$	
12. $O^- + O_2 \rightarrow O_2^- + O$	$2.5 \times 10^{-14}\ cm^3/sec$ at $E/p = 20$ V/cm torr; $3.4 \times 10^{-12}\ cm^3/sec$ at $E/p = 45$ V/cm torr	
13. $O^- + O_3 \rightarrow O_3^- + O$	$5.3 \times 10^{-10}\ cm^3/sec$	
14. $O^- + 2O_2 \rightarrow O_3^- + O_2$	$1.0 \pm 0.2 \times 10^{-30}\ cm^6/sec$	
15. $O_2^- + O \rightarrow O^- + O_2$	$5 \times 10^{-10}\ cm^3/sec$	
16. $O_2^- + O_2 \rightarrow O_3^- + O$		$<10^{-18}$
17. $O_2^- + O_3 \rightarrow O_3^- + O_2$	$4.0 \times 10^{-10}\ cm^3/sec$	
18. $O_2^- + 2O_2 \rightarrow O_4^- + O_2$	$3 \times 10^{-31}\ cm^6/sec$	

Table 1.3, continued

Reaction	k	$\sigma_{max}(cm^2)$
Charge Transfer, continued		
19. $O_3^- + O_2 \rightarrow O_2^- + O_3$		4×10^{-17}
20. $O_4^- + O \rightarrow O_3^- + O_2$	4×10^{-10} cm^3/sec	
21. $O_4^- + O_2 \rightarrow O_2^- + 2\,O_2$	6×10^{-15} cm^3/sec	
Detachment		
22. $O^- + O \rightarrow O_2 + e$	3.0×10^{-10} cm^3/sec	
23. $O^- + O_2 \rightarrow O + O_2 + e$		7×10^{-16}
24. $O^- + O_2(^1\Delta_g) \rightarrow O_3 + e$	$\sim 3 \times 10^{-10}$ cm^3/sec	
25. $O_2^- + O \rightarrow O_3 + e$	5.0×10^{-10} cm^3/sec	
26. $O_2^- + O_2 \rightarrow 2\,O_2 + e$		7×10^{-16}
27. $O_2^- + O_2(^1\Delta_g) \rightarrow 2\,O_2 + e$	$\sim 2 \times 10^{-10}$ cm^3/sec	
Electron-Ion Recombination		
28. $e + \begin{Bmatrix} O^+ \\ O_2^+ \\ O_3^+ \\ O_4^+ \end{Bmatrix} \rightarrow \begin{Bmatrix} O \\ 2\,O \\ O + O_2 \\ 2\,O_2 \end{Bmatrix}$	$<10^{-7}$ cm^3/sec	
Ion-Ion Recombination		
29. $\begin{Bmatrix} O^- \\ O_2^- \\ O_3^- \\ O_4^- \end{Bmatrix} + \begin{Bmatrix} O^+ \\ O_2^+ \\ O_3^+ \\ O_4^+ \end{Bmatrix} \rightarrow \begin{Bmatrix} O \\ O_2 \end{Bmatrix}$	$\sim 10^{-7}$ cm^3/sec	
Atom Recombination		
30. $2\,O + O_2 \rightarrow 2\,O_2$	2.3×10^{-33} cm^6/sec	
31. $3\,O \rightarrow O + O_2$	1.5×10^{-34} cm^6/sec	
32. $O + 2\,O_2 \rightarrow O_2 + O_2$	1.9×10^{-35} exp $(2100/RT)$ cm^6/sec	
33. $O + O_3 \rightarrow 2\,O_2$	2.0×10^{-11} exp $(-4790/RT)$ cm^3/sec	
34. $O \xrightarrow{\text{wall}} \frac{1}{2}\,O_2$	$\gamma = 1.6 \times 10^{-4}$ to 1.4×10^{-2} (T = 20-600°C)	

[a]Reprinted by permission of John Wiley & Sons, Inc.

or

$$\frac{v_i}{p} = \frac{D_{ei}}{(p\Lambda)^2}$$

where D_{ei} = electron diffusivity at 1.0 Torr
v_i/p = ionization constant $k_i = v_i(RT)p$ and is a function of E/p

Figure 1-4 shows the relationship of v_i/p vs E/p.

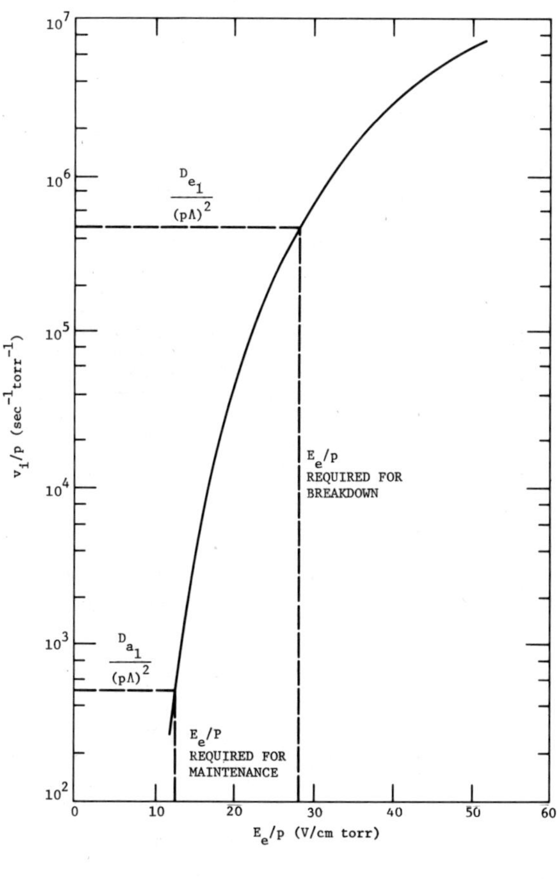

Figure 1-4. The breakdown and maintenance values of E_e/p on a plot of the reduced ionization frequency versus E_e/p for hydrogen (Hollahan and Bell, 1974).

PHYSICAL DESCRIPTION OF THE CORONA PHENOMENON

Corona is a gas-phase phenomenon. Corona produces free radicals in a mixture of gases. In addition to reacting in the gas phase the radicals can also interact with molecules of a liquid or finely dispersed solid exposed to them.

Figure 1-5 ia a typical voltage-current plot for a positive needle-to-plane configuration in atmospheric air as driven by a dc power supply (Gallo, 1977). Analysis of this curve and each of the distinctive regions allows description of the phenomenon.

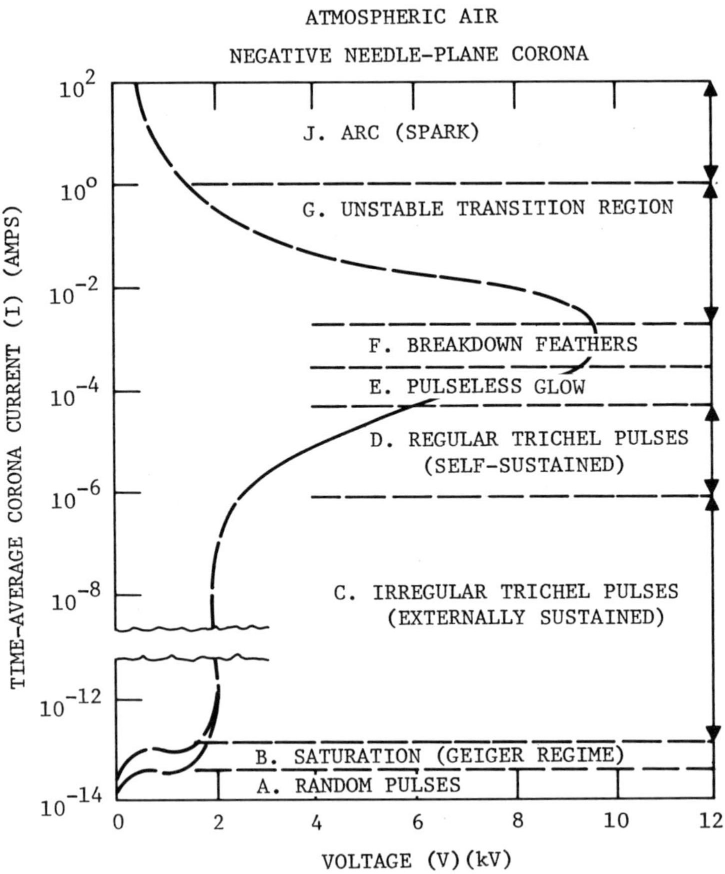

Figure 1-5. Schematic of current-voltage relation for air around atmospheric pressure (Gallo, 1977).

1. *Region A*. The phenomenon of corona starts with a few stray electrons always present in a gas because of the effect of cosmic and other background radiation. Originally, the electric field is low enough that the electrons do not have and cannot acquire the necessary energy to produce further ionization. If from an external source additional electron and/or ionization is introduced, the current will increase.

2. *Region B*. As the voltage is further increased, the current reaches a saturation level where all the charges externally generated will be swept out of the gap. Saturation curves will vary with the rate of external charge generation. This region is known as the Geiger regime.

3. *Region C.* At still higher voltages, the energy acquired by the electron becomes sufficient to ionize the gas by collision, forming additional electrons, leading to additional ionization. An avalanche occurs and the current rises rapidly. This is the Townsend avalanche region. Irregular current pulses, called onset streamers or burst pulses, occur.

4. *Region D*. As the voltage increases, these pulses become self-sustaining, i.e., no longer dependent on external ionization.

5. *Region E*. As the voltage is further increased, a steady state occurs as charges accumulate and continue the self-sustaining reaction. The anode starts to glow. This is the corona glow region. Figure 1-6 provides a more detailed depiction of the macroscopic phenomena of corona glow; as electrons accelerate toward the positive electrode, they leave a trail of positive ions. Energetic photons ionize atoms and molecules in their path. Electrons following the channels of positive ions toward the cathode create an avalanche and the filamentary structure known as onset streamers. Eventually, the channels are dispersed due to charge repulsion and the positive ions drift away from the anode while the negative ions move toward the cathode.

6. *Region F*. At higher voltage, breakdown streamers appear. These are stronger than the onset streamers previously encountered.

7. *Region G*. When the voltage reaches a critical value, ionization occurs across the entire gap, creating an unstable transition region.

8. *Region H*. The unstable transition discharge evolves suddenly into a destructive high current arc or spark.

MATHEMATICAL DESCRIPTION OF THE CORONA PHENOMENON IONIZATION COEFFICIENT

An electron will ionize on collision with an atom when its kinetic energy gained prior to collision is equal to the ionization eV_i of the atom. If λ_i is the free path in the direction of the field, then

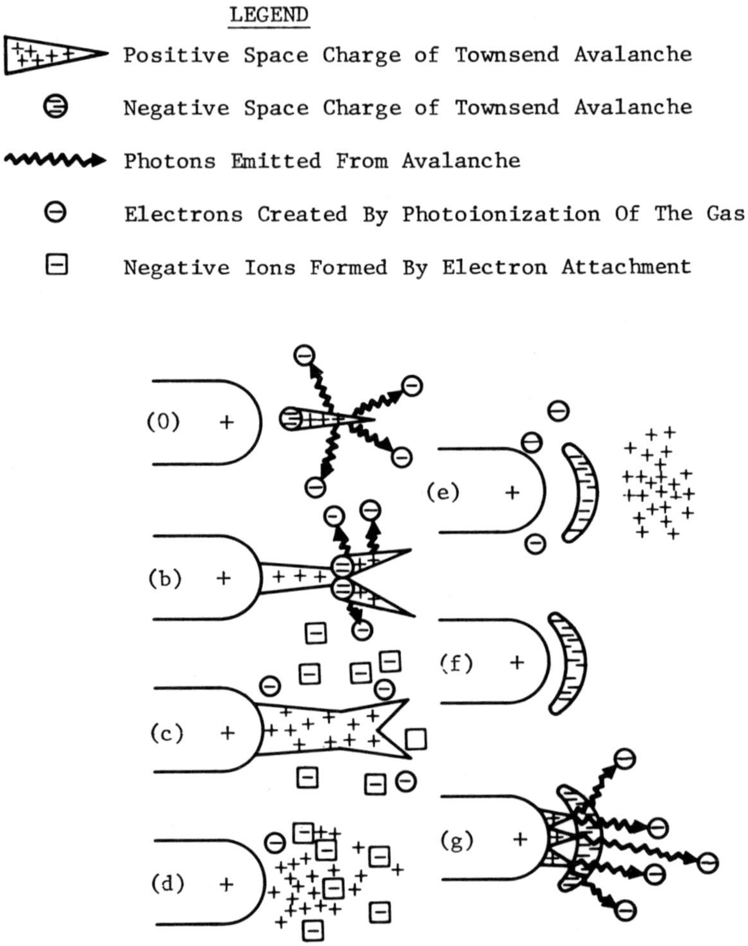

Figure 1-6. Schematic diagrams showing the streamer-glow transition in positive needle coronas (Gallo, 1977): (a) Initial Townsend ionization (burst pulse) avalanche with photons created therein and subsequent photoionization of the surrounding gas; (b) Streamer growth by additional ionization avalanches initiated by photoelectrons in strong field regions; (c) Attenuation of the streamer and the formation of negative ions in the weak field regions by electron attachment of the photoelectrons; (d) Drifting positive and negative ions; (e) Formation of negative ion space charge cloud; (f) Creation of locally high field between the negative ion charge cloud and needle tip; (g) Townsend ionization avalanches (denoted by T) initiated by field stripping of electrons from the negative ions approaching the needle tip. Note that the photons created by the Townsend avalanches (T) photoionize the gas and create electrons, which attach to gas molecules to form negative ions drifting in the weak gap field, thus continually replenishing the negative ion cloud being depleted by field stripping at needle side of the cloud.

$$eE\ \lambda_i \geqslant eV_i$$

$$\lambda_i = V_i/E$$

The number of free paths, m, which is equal to or greater than x, is

$$n = n_o\ {-x/\bar{\lambda}}$$

where $\bar{\lambda}$ is the mean free path.

Defining n_E as the number of the paths or number of collision per unit length along E, and $\bar{\lambda}_E$ as the corresponding mean free path:

$$\bar{\lambda}_E = \frac{1}{n_E}$$

The relation between $\bar{\lambda}$ and $\bar{\lambda}_E$ depends on the drift velocity, v, of the electron that is its speed along E.

We can now write

$$\frac{\bar{v}}{\lambda} = \frac{V}{\bar{\lambda}_E}$$

The number of ionizing collisions arising from the moving electron along E per unit length is

$$\alpha = n_E e^{-\lambda_{ie}/\bar{\lambda}_E} = n_E e^{-\lambda_i/\bar{\lambda}}$$

$$= \frac{1}{\bar{\lambda}_E}\ e^{-\lambda_i/\bar{\lambda}}$$

According to the above equation, $\bar{\lambda}$ is proportional to T/p or $1/\bar{\lambda} = Ap$, where A is a constant. Substituting,

$$\alpha = Ape^{-V_i/E\bar{\lambda}}$$

$$= Ape^{-ApV_i/E} = Ape^{-Bp/E}$$

Since $\bar{v} \approx v$ and $B = AV_i$,

$$\frac{\alpha}{p} = A_e^{-Bp/E}$$

The important finding is that α/p is a function of E/p.

Townsend's Discharge

Since the Townsend Region is of particular interest to this book, we will review in more detail the physics and mathematics of the Townsend Discharge.

Consider two plane electrodes (anode and cathode), separated by a distance, d, and plate A being V volts above plate B. If n_o electrons start at the plane cathode, we expect n electrons at the opposite plane, the anode, where n is given by

$$n = n_o e^{\alpha d}$$

Multiplying both sides by the charge of the electron e gives

$$en = en_o e^{\alpha d}$$

The current measured by placing the electrodes at various distance is

$$i = i_o e^{\alpha d}$$

The distance from the cathode in which ionization takes place is

$$d = \frac{V}{E}$$

$$\ln \frac{i}{i_o} = \alpha d = \alpha \frac{V_i}{E}$$

Hence, after the onset of ionization by electron collision, the current increases exponentially with electrode separation at a constant, E, and constant, E/p, where p is pressure (see Figure 1-7) and E is the field.

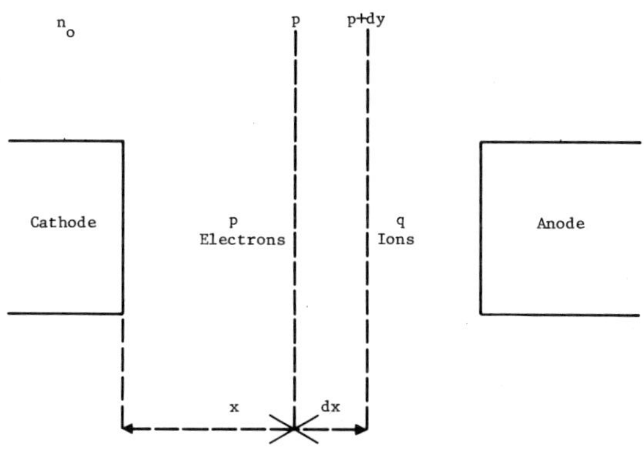

Figure 1-7. Electrodes configuration (Hollahan and Bell, 1974).

If p and q ionization events take place to the left and right of the plane x, as shown in Figure 1-7, the total number of electrons created in the tube of unit cross section per second is

$$n = n_o + p + q$$

Within the slab dx, these electrons produce by collision a number of electrons and ions equal to

$$(n_o + p)\ \lambda dx$$

The number of q ions entering the slab produce additional electrons and positive ions equal to

$$q\beta dx$$

where β = ionization coefficient per unit length.

The total number of ionization events between x and x + dx is

$$(n_o + p)\,\alpha\,dx + q\beta dx$$

which is equal to dp:

$$dp = [(n_o + p)\ \alpha + q\beta]\ dx$$

and the number of electrons leaving the plane x + dx is given by

$$p + dp + n_o$$

The result is

$$dp = [\ \lambda(n_o + p) + \beta(n - n_o - p)]\ dx$$

Intergrating this, yields

$$n = n_o \frac{(\alpha-\beta)e^{(\alpha-\beta)d}}{\alpha-\beta e^{(\alpha-\beta)d}}$$

or, in terms of current:

$$\frac{i}{i_o} = \frac{(\alpha-\beta)e^{(\alpha-\beta)d}}{\alpha-\beta e^{(\alpha-\beta)d}}$$

The condition for the current to be sustained is

$$\alpha - \beta e^{(\alpha-\beta)d} = 0$$

or

$$\frac{\alpha}{\beta} = \frac{e^{\alpha d}}{e^{\beta d}}$$

For a>>β,

$$\frac{n}{n_o} = \frac{i}{i_o} = \frac{e^{\alpha d}}{(1-\beta/\alpha)e^{\alpha d}}$$

and the breakdown criterion becomes

$$\frac{\alpha}{\beta} = e^{\alpha d}$$

Ozone Generation

One of the best understood and documented corona phenomenon is the generation of ozone (Gallo and Castle, 1978).

A number of historical observations (Table 1-4) form the basis for any theory of ozone generation. The physical model is dc Corona from a metallic wire within a coaxial metal cylinder with axial air flow.

Castle et al. (1969) developed the following empirical relationship:

$$[O_3]'/I = 220\,(E_{av}v) \quad (1)$$

where E_{av} = average electrical field across the ionization sheath.

Table 1-4. Ozone Generation–Empirical Observations

Formulation	Observations
$[O_3]' \propto I$	The ozone generation rate ($[O_3]'$) is directly proportional to the corona current (I) when all other parameters are held constant.
$[O_3]' = [O_3]Q$	The ozone generation rate is equal to the product of the ozone concentration ($[O_3]'$) and the air flowrate (Q).
$\frac{[O_3]'}{I} = f_+(r)$	The ozone generation rate per unit current increases in a monotone fashion (f_+) with the corona wire radius (r), all other parameters being held constant.
$\frac{[O_3]'}{I} = f_-(T)$	The ozone generation rate per unit current decreases in a monotone fashion (f_-) with the temperature (T) of the corona wire.
$[O_3]'\text{-ve} = K[O_3]'\text{+ ve}$	The ozone generation rate per unit current in the positive glow mode (+ve) is much less ($k \approx 5\text{-}8$) than is the negative glow mode (–ve).
$[O_3]' \propto P_i$	The ozone generation rate is directly proportional to the electrical power dissipated in the ionization sheath surrounding the corona wire.

Gallo et al. developed a simplified explicative theory as follows:

The ozone generation rate is proportional to the power dissipated in the ionization sheet surrounding the corona wire:

$$[O_3]' \propto P_i \tag{2}$$

The power dissipation is equal to the voltage across the ionization sheath and the corona current:

$$P_i = V_i I \tag{3}$$

Assuming that the voltage, V_i, is proportional to the threshold voltage, V_o, and combining the equations above, we obtain:

$$[O_3]' \propto P_i \propto V_o I \tag{4}$$

The relationship between corona onset voltage, V_o, and ozone generation rate per unit current ($[O_3]'/I$), is linear but does not pass through the zero intercept, suggesting a modification to the above equation:

$$[O_3]'I \propto (V_o - D) \tag{5}$$

The negative contribution of D has been attributed to ozone destruction and to the voltage drop across the corona drift region at threshold, among others.

Gallo and Castle (1978) developed two functional relationships in agreement with the experimental data, based on Equations 1 and 4 above, as follows:

$$\begin{aligned} [O_3]'/I &\propto (E_{av}r) = 30r + 4.5\sqrt{r} \\ [O_3]' &\propto V_o = (30r + 9\sqrt{r})\ \ln b/r \end{aligned} \tag{6}$$

where b = radius of the outside cylinder.

Ozone is produced by the reaction of O atoms with molecular oxygen. Ozone can be produced by ultraviolet radiation and electrical discharges (American Chemical Society, 1959; Murray and Orr, 1975):

$$O_2 + hv \longrightarrow 2O$$
$$O + O_2 \longrightarrow O_3$$

Ozone itself is decomposed by photolysis, electron impact or thermally:

$$O_3 \xrightarrow{\Delta} O + O_2$$
$$O + O_3 \longrightarrow 2O_2$$

Looking in more detail at the ozone produced by electrical discharge, we see that it is a chain reaction sustained as long as the applied voltage is maintained:

$$O_2 + e^{-*} \longrightarrow O_2 + 2e^{*-}$$
$$\longrightarrow O_2^{*}$$
$$O_2^{+} + e^{-} \longrightarrow O_2^{*}$$
$$O_2^{*} \longrightarrow 2O$$
$$O + O_2 \longrightarrow O_3$$

where * = energy >>> KT.

CONFIGURATIONS AND/OR DEVICES

On a more pragmatic basis, the devices used for generating a plasma environment can be reviewed in terms of Figure 1-8 (Blausten, 1969):

1. the source of electrical power
2. the power being delivered by a coupling mechanism to
3. a plasma environment

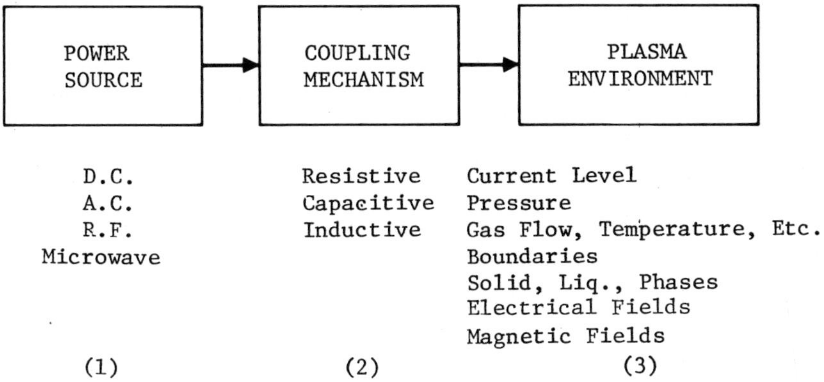

Figure 1-8. Elements of an electrical discharge device (Blausten, 1969).

Power Source

The power source can be a direct current, an alternating current, a radio frequency or a microwave.

Coupling Mechanisms

Coupling mechanisms can be resistive, capacitive or inductive in nature. It is a resistive for those devices where the electrodes are in direct contact with the ionized gas or plasma environment.

In inductively or capacitively coupled devices, electrodes are not in direct contact with the plasma. The electric field is generated or induced by a changing magnetic field.

Discharge devices coupled by capacitance are further characterized by the fact that the electrode is separated from the plasma by a dielectric barrier, i.e., a nonconducting material.

Plasma Environment

The properties of the plasma environment can be readily altered by a variety of external factors. They include gas pressure; gas flow rate; inlet temperature; frequency, voltage, current and duration of the input power; and the presence of external magnetic fields.

It is then possible to control reaction rates in the plasma region by manipulating the above variables in their many combinations.

SPECTRUM OF APPLICATIONS

We will address two narrow domains in the field of plasma chemistry, namely:

1. surface modification
2. organic synthesis

However, it is important to recognize the immense potential of plasma in research and industry. To provide this flavor, we will rapidly review the entire spectrum of plasma applications, as shown in Table 1-5.

Nonimpact Printing

Nonimpact printing is based on electrostatic forces used to modify liquid surfaces and produce images of ink droplets electrostatically detected into a meaningful image. Electrical discharges delivered by metal styli are used to place

Table 1-5. Taxonomy of Applications

Chemical Synthesis	Reaction of atomic hydrogen, oxygen, nitrogen with organic compounds *Cis-trans*-isomerization Migration of substituents Rearrangement of cyclic compounds Eliminations Cyclizations Bimolecular reactions
Surface Modification	Polymers – wettability/adhesion molecular weight change composition change shrinkproofing grafting Metals – nitriding coating carbiding siliciding etching Semiconductors – plasma etching photoresist stripping surface cleaning
Electrostatic Imaging	
Nonimpact Printing	
Air Pollution Control	Electrostatic precipitation Water pollution control (O_3)
Chemical Physical Analyses	Ashing Removal of organic matter Reaction to gaseous products Preconcentration Cleaning
Dielectrophoresis	
Electrostatic Separation	Mineral beneficiation Purification of food Sorting of reusables Electrostatic sizing
Electrostatic Motors	
Electrostatic Generators	
Biological Applications	Low-temperature ashing Sterilization Improved cell attachment to surfaces

a pattern of charges into a dielectric surface. The charge patterns are delivered by electrophoretic toners.

Because of its advantages (Table 1-6), nonimpact printing is being utilized in many computer printing applications.

Table 1-6. Comparison of Impact and Nonimpact Printing (Moore, 1973)

	Nonimpact	Impact
Print Quality	**Dot Matrix Characters**	**Full Characters (Imperfect Alignment)**
Multiple Copies	No	Up to 6
Speical Forms	Depends on technology	No
Print Speed	5000 lpm[a] (30,000 lpm)	200 lpm with 48 characters
Noise Level	Inherently low	High
Font Size	Limited only by control electronics	Mechanically limited (affects print speed)
Font Selection	Logic controlled	Required mechanical change
Reliability	Potentially better than impact	

[a]lpm = lines per minute.

Electrostatic Separation

Solids in air, liquid, medium or gases can be separated from each other based on their electrostatic properties.

Electrostatic separation has been applied to the following:

- benefication of ores,
- purification of ores,
- sorting of reusable wastes, and
- sorting of products according to size and/or shape.

Of the numerous industrial applications, some of the most important are listed in Table 1-7. The principle is the same for all applications. The utilization of forces acting on charged bodies is an electric field. It is estimated that 14 million tons of products are separated electrostatically each year worldwide. Mineral beneficiation accounts for 90% of the volume produced.

Table 1-7. Industrial Separations by Electrostatics

Arsenopyrite-Feldspathic Gangue	Mica-Feldspars/Silicates
Asbestos-Silicates	Molybdenite-Silicates
Barite-Silicates	Monazite-Beach Sands
Chromite-Ilmenite-Magnetite-Monazite	Nickel-Copper Ores-Metasilicates
Coal-Pyrite	Rutile-Beach Sands
Coal-Shale	Scheelite-Pyrite
Cobalt-Silver-Silicates	Silicon Carbide-Alumina/Silicates
Coke-Iron	Spodumene-Cassiterite
Copper Ore-Silicate	Stibnite-Silicates
Copper Wire-Insulation	Wire-Thermoplastics/Rubber
Diamonds-Silica	Wolframite-Pyrite
Feldspar-Mica Gangue	Zircon-Beach Sands
Feldspar-Quartz	Bark-Sand
Fluorite-Silicates	Barley-Rodent Excrement
Fly Ash-Carbon	Cocoa Beans-Shell
Gold/Platinoids-Beach Sand	Cotton Seeds-Stems
Gold/Platinum-Jewelry Sweeps	Movie Film-Paper
Graphite-Silicates	Nut Meats-Shells
Halite-Sylvite	Peanuts-Shells
Ilmenite-Garnet	Plastic-Lint
Ilmenite-Beach Sand	Polyvinyls-Polyesters
Iron (specular hematite)-Silicates	Rice-Rodent Excrement
Kaolin-Iron Contamination	Seeds-Foreign Material
Kyanite-Rutile and Iron Gangues	Soap-Detergent
Limestone-Silicates	Soybeans-Rodent Excrement
Magnetite-Silicates/Rutile	Walnuts-Shells
	Wheat-Garlic Seeds

Semiconductors

Plasmas play an important role in modern manufacturing and processing of semiconductors (LFE Corp.: Circuit Manufacturing, 1971; Poulsen, 1976; Franson, 1974; Yang, et al., 1975; Irving, 1968; O'Kane, 1974). They are used today in plasma cleaning, photoresist stripping, etching and silicon nitride deposition (Figure 1-9).

Plasma Cleaning

The use of fluorine-containing plasma can be an extensive technique for cleaning prior to metal evaporation in single and multilayer metalization schemes; deposition of passivation glass; wire bonding; and gold backing.

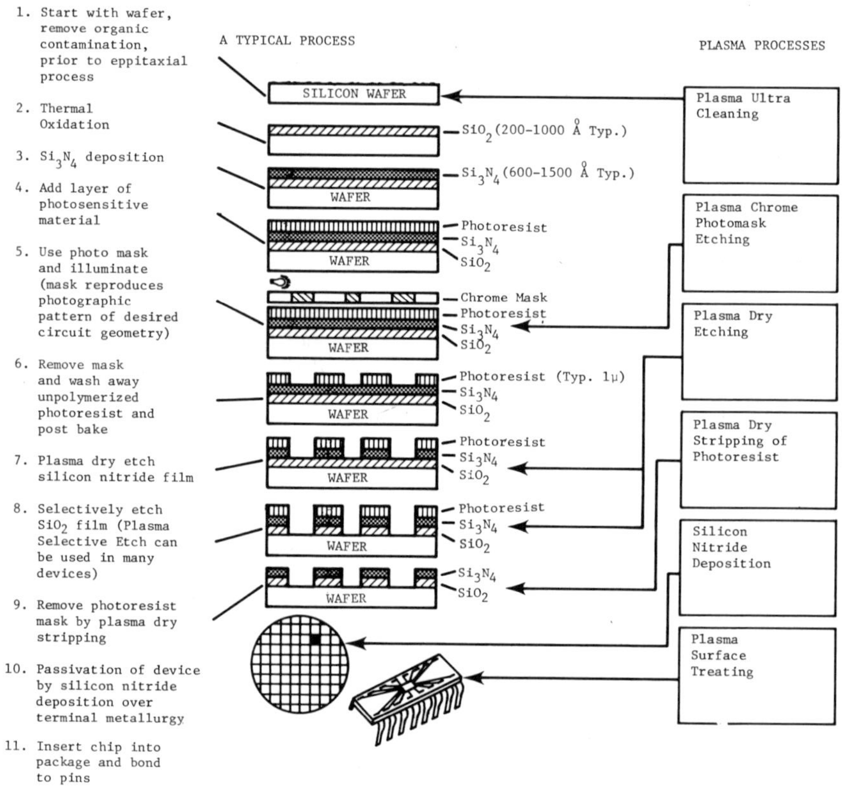

Figure 1-9. Microcircuit fabrication (LFE Corporation).

Photoresist Stripping

Using oxygen plasma for the removal of photoresist films was introduced in 1968 by Irving. The presence of trace amounts of photoresist is called scumming. Engineers have tried to solve this problem by increasing developing and/or rinse time, varying the prebake time and/or temperature, etc. A few minutes of exposure to an oxygen plasma removes approximately 1000 Å of potential scum in the window area without otherwise affecting the surface.

Plasma Etching

The plasmas used in plasma etching are ionized gases containing reactive ions, free electrons and radicals. The plasma are characterized by low temperature (50-250°C) and low pressure (13-400 Pa) and etching rates on the order of 50-500 cm^3/min.

The chemical vs plasma etching process is compared in Figure 1-10, while advantages and disadvantages are presented in Table 1-8. Some relative etching rates are given in Table 1-9.

Biological Applications

Three biological applications (Smith, 1973; Smith et al., 1975; Gleit, 1963; Mulford, Hollahan) of low-temperature plasma are of interest:

1. low-temperature dry ashing of biological segments;
2. sterilization, and
3. improved cell attachment to surfaces.

Low-Temperature Ashing

Plasma ashing has been used for removing carbon and organics before particulate analysis (recovery of airborne particles, or particulate residues chemically extracted from steel, of calcareous or siliceous skeletical material in biology) and for microincineration (careful burning of biological specimens to demonstrate their structure by ash patterns, in three and two dimensions) applied to a large number of biological, biomedical and cellulosic specimens (Tsuji, 1977).

Ashing of biological materials is essential to allow for subsequent elemental analysis. Basically, three methods are available:

1. wet ashing, principally with mixtures of hot mineral acids;
2. application of heat directly to the sample; and
3. excited oxygen plasma treatment at low temperature.

The low-temperature plasma volatilizes the organic matrix (H_2, C, N and O) leaving the inorganic structure.

The rate of sample ashing on exposure to the plasma is determined by:

1. available surface area for excited oxygen to contact;
2. concentration of excited oxygen available for reaction;
3. temperature, which affects the rate of chemical reaction; and
4. the chemical nature of the sample material.

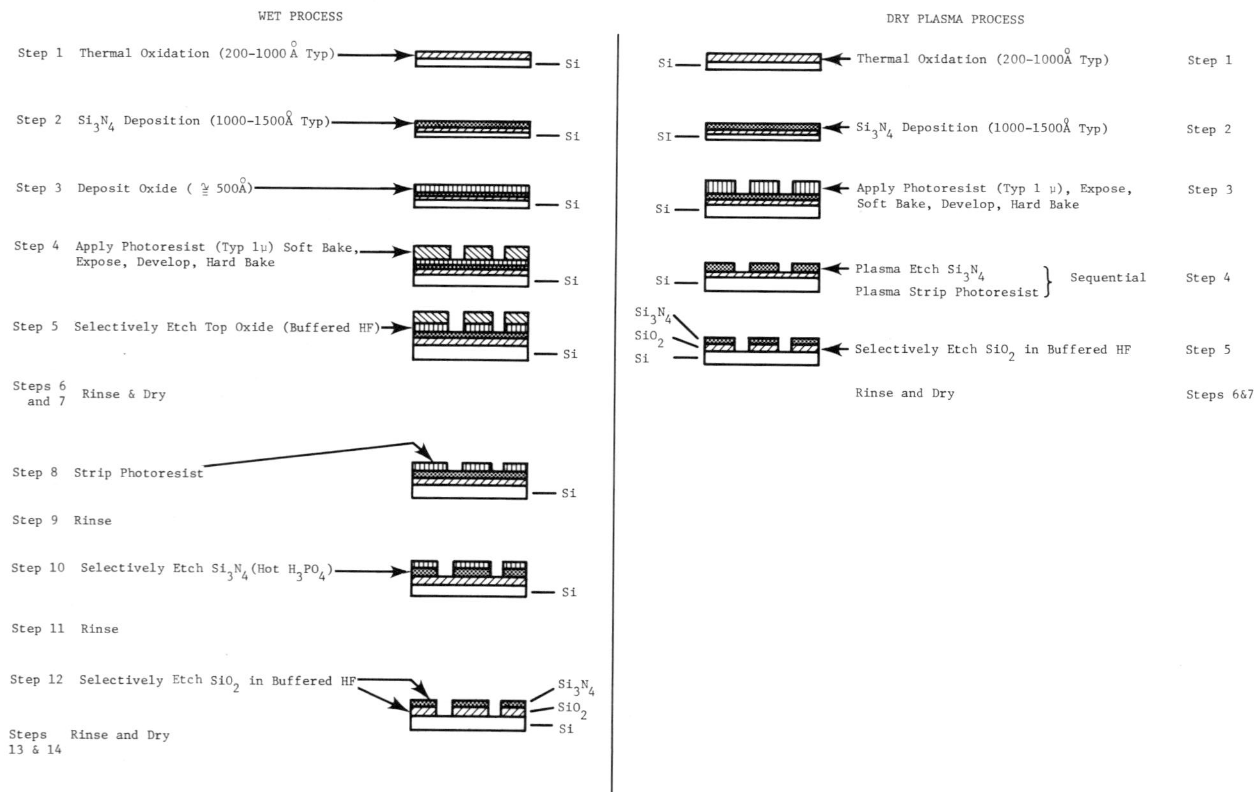

Figure 1-10. Comparison of wet and dry plasma processes (LFE Corporation).

Table 1-8. Some Advantages and Disadvantages of Plasma Etching and Stripping (LFE Corporation)

Advantages

1. Avoids use of expensive polluting chemicals and solvents.
2. Fast and inexpensive processes.
3. Simplifies processes and production line.
4. Improves yield since cleaned with little wafer handling.
5. Improves pattern resolution and tolerances.
6. Tightens product specifications.

Disadvantages

1. Etching equipment more expensive and complicated.
2. Etching processes not yet well understood.

Table 1-9. Relative Etching Rates[a] of Common Microelectronic Materials (LFE Corporation)

In CF_4 + 4% O_2 Plasma[b]	
Silicon nitride	100
Silicon (111)	690
Silicon (polycrystalline)	990
Silicon dioxide (thermal)	40
Silicon dioxide (CVD)	120
Tungsten	100
Molybdenum	100
Titanium	100
Tantalum	100
Aluminum	0
GaAsP	0
Waycoat IC photoresist	20
AZ1350J photoresist	40
In CL_2 += 50% O_2 Plasma[c]	
Chromium	20
In $C_2Cl_2F_4$ Plasma[c]	
Gold	100

[a]The etch rates are in units of Å/min.
[b]With etch tunnel.
[c]Without etch tunnel.

Sterilization

The sterilization of surfaces contaminated by microorganisms has been achieved in 10-15 minutes in a cool (25-75°C) ionized gas plasma at about 1.0 torr in oxygen atmospheres, or with rare gases, nitrogen, hydrogen, etc. The single-cell organisms are killed on the basis of a dry gas/solid process.

Improved Cell Attachment to Surface

Using an inert gas glow discharge, a vast improvement in tissue cell culture attachment to the surface is achieved, providing a more reliable cell substrate interface while doubling the cell plating efficiency. A three-minute treatment in a cool (30-70°) argon plasma gives results as in Figure 1-11 comparing the degree of cell adhesions to various surfaces vs shear stress.

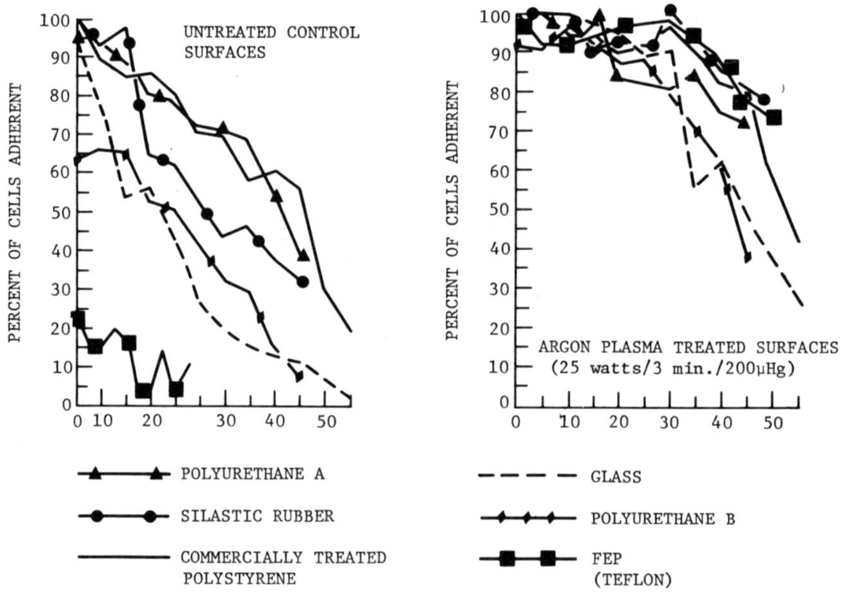

Figure 1-11. Adhesion of 3T3 fibroblasts. (a) Shear stress (dynes/cm^2). (b) Shear stress (dynes/cm^2). (Smith et al., 1975).

Metal Surfaces

The treatment of plastic films is widely practiced to promote adhesion of inks, coatings and adhesives. A number of patents have been issued, but few applications of corona to metal surfaces can be found in the industry.

One such interesting application is the treatment by electrostatic discharges of as-rolled aluminum foil (Stillwagon and Butcher, 1971).

Inline electrostatic treatment of as-rolled foil was studied in combination with various converting operations, such as laminating, lacquering and extrusion coating. Tests indicated improved surface adhesion of nitrocellulose inks to the special waste-coated, electrostatically treated foil. The same was shown for the adhesion of water-based coatings and the adhesion of polyethylene extrusion coatings. The improvement in adhesion disappears over time and seems to be associated with residual rolling lubricant or organic contaminants on the surface.

Theories have been proposed to explain the phenomenon. Oxidation of the organic materials by ozone would render the surface polar and more reactive to coatings. Another theory assumes reorientation of polar groups on the surface.

Another application is surface nitriding, associated with the fabrication of conductors. Silicon nitride, because of its high density, can mask against diffusants--unintentional or used in semiconductor processing. Industrial operations at low pressure (few torr) and temperature ($<350°C$) have achieved a silicon nitride deposition by plasma means at a rate exceeding 1000 Å/min.

Other Surface Modifications

It is often desirable to improve the glossy appearance or reflectivity of materials. This can be achieved by depositing thin aluminum or other metallic coatings on plastics, glass or pretreated metals. Such coatings are not normally resistant to normal atmosphere and must be covered by an additional layer or topcoat.

Plasma polymerization in the vacuum chamber allows an organic protective layer to be overlaid on an aluminum coating (Dittmer, 1978). SiO_2 has been used in the manner as a topcoat with excellent chemical resistance.

Sputter ion plating (Dugdale and Coad, 1978) is another method for depositing material from an appropriate cathode (target) by the action of ion bombardment (sputtering) on substances in a vacuum chamber. An inert gas, usually argon at a pressure in a range of 10^{-5}-10^{-4} bar is the environment. Temperatures above 200°C are usually used. Typical deposition rates and specific energy consumption are given in Table 1-10 for metals of interest. A chart of adherent coating substance combinations successfully deposited is given in Table 1-11.

The main advantages of ion sputter coatings are absence of porosity, high adherence, compositional control and deposition at low and moderate temperatures. Major industrial applications are for wear resistance, corrosion resistance, joining and decorative uses. The automotive industry appears to be the stronger market.

Electrostatic Imaging

Electrostatic imaging is a process in which electrical charges are deposited on a surface in a prearrangement pattern. The pattern is sequentially developed to produce a visual image. The essential steps of the process are:

Table 1-10. Deposition Rates for Dense Coatings and Specific Energy Consumption Data Obtained in Early Trials (Dugdale and Coad, 1978)

Coating	Deposition Rate (μm/hr)	Specific Energy Consumption (eV/atom)	(kWh/g)
Al	1	1800	2
Ni	1	1070	0.5
Ti	1	2470	1.4
Cr	0.5	1350	0.7
Mo	5	1430	0.4
W	0.5	1250	0.2
St/St	2	1440	0.7

- surface charging
- latent image formation
- image development
- image transfer, and
- fixing.

Xerography is one such process used for duplicating material on paper.

Optical Surfaces

Optical surfaces may often be contaminated by low-level organics, films of polymers, chemisorbed species and microorganisms. These species degrade the optical properties of the surfaces.

Atomic oxygen generated in a cool plasma oxidizes away the contaminant films in a dry gas/solid process (Gillette et al., 1970).

Low-temperature plasma perfluorinated compounds generate a polymer that can be used as coating for optical surfaces such as alkali halide crystals. This film, which is pin-hole free and thermally stable, protects the surface from moisture while retaining optical clarity (Hollahan and Weydeven, 1974).

Environmental Protection

The need for environmental protection has stimulated the invention of a number of ingenious uses of plasma. Three of the most important, which will be described below, are:

1. electrostatic precipitation of particulates,
2. ozone treatment of water, and
3. destruction of organic compounds.

Table 1-11. Adherent Coating/Substrate Combinations (o) Formed by Sputter Ion Plating (thickness range 2-150 μm) and Some Combinations (x) That Formed Brittle Intermetallic Phases (Dugdale and Coad, 1978)

Coating \ Substrate	Al	Aluminized Kapton	Al_2O_3	Cast Iron	Cr	Cu	Fecralloy	Glass	IN 100 (Stellite)	Mo	Ni	Nilo K	Porcelain	SiC	Si_3N_4	SiO_2	Stainless Steel	Steels (various ferritic)	Ta	Ti	W
Ag																	o				
Al			o			o											o	o			
Al_2O_3						o					o						o				
Al-N											o										
C										o							o			o	
CoCrAlY									o		o						o	o			
Cr										o	o							o			
CrC	o									o	o						o	o	o		o
Cu	x		o	o						o	o					o	o	o	o	o	
Fecralloy	o			o		o		o		o	o			o	o		o	o		o	
Mo	x		o	o		x	o			o	o	o				o	o	o	o	o	o
Nb											o	o									
Ni	x	o	o	o		o	o	o					o		o	o			o	o	o
Stainless Steel	o										o						o	o			
Ti	x					o					o	o					o	o			
Ti-C	x			o						o	o						o	o	o	o	o
Ti-C-N																		o		o	
Ti-N	o		o		o	o	o			o	o		o			o	o	o	o	o	o
TiO_2											o							o		o	
Ti-O-N			o								o						o	o			
Ti-Ta-C											o						o	o		o	
W											o						o	o		o	
W-C				o						o								o	o	o	o
WC (6% Co)																		o			
WCoCrNiC (X40)											o		o								
Zr-Ti-N				o						o	o						o	o	o	o	

Electrostatic Precipitation

Electrostatic precipitation originated with Frederick Cottrell in 1901. The first such precipitator was introduced in 1923 by the Detroit Edison Company to trap fly ash generated during the combustion of coal.

Today there are an estimated 4000 industrial precipitators in use in the U.S., distributed as follows:

- Power industry, 60%
- Steel industry, 10%
- Cement industry, 10%
- Paper industry, 7%
- Nonferrous metals industry, 7%
- Chemical industry, 6%

Two types are in use: duct type and tubular arrangement. If sufficient potential is applied between the discharge and the collecting electrodes, a corona discharge will develop between the electrodes forcing the migration of ions. In the path between the electrodes, the ions attached themselves to aerosol particles. The charged particles are driven towards the collecting surfaces where they adhere.

Water Pollution Control with Ozone

Potential applications of ozone are too numerous and too diversified to describe in detail. Listed below are a few of the specific applications where ozone is being commercially employed or extensively studied (Rosen, 1977; Drahusink and Rosen, 1977; Rosen, 1976):

1. Municipal Wastewater Treatment:
    - Disinfection with inherent dissolved oxygen elevation
    - Polishing
    - Toxicity reduction
    - Reuse
2. Municipal Potable Water Treatment
    - Disinfection
    - Reduction of trihalomethanes through precursor oxidation
    - Oxidation for removal of:
        - Color
        - Taste
        - Odor
        - Manganese and iron
        - Other organic and inorganic contaminants
3. Industrial Water and Wastewater
    - Oxidation for removal of:
        - Color
        - Phenols
        - Cyanides
        - Other oxidizable organic and inorganic pollutants
        - Recycle and reuse
        - Disinfection and biocide
4. Industrial Process
    - Bleaching
    - Chemical synthesis
5. Air Pollution
    - Odor control
    - Other stack gas emissions control

Destruction of Organic Compounds

Microwave plasma has been used by scientists at Lockhed Missiles and Space Company to break down concentrated toxic chemicals.

PCB (Aroclor 1242 and 1252), methyl bromide, malathion, phenyl muric acetate (PMA), kepone, red dye smoke and diisopropyl methylphosphonate are among the chemicals treated.

A first cost estimate is that a unit with a 100-lb/hr capacity would entail an operating cost of about 25 ¢/lb of material treated.

Chemical/Physical Analyses

Plasmas are used in a sample preparation to ash a sample prior to elemental analysis. This method has many advantages (including lower loss of volatile elements) over the traditional chemical ashing or the temperature ashing methods. Applications to coal analyses and air pollution (particulate) samples analyses can be found in the literature (Kometani et al., 1972; Kneip et al., 1970; List et al., 1968; Hauser, 1977; Gluskoter, 1976).

Plasmas are promising as analytical tools (Fairless, 1978) for multielements analysis. Table 1-12 compares dc plasma analysis with more traditional and established methods. It is shown that plasma analysis is competitive with the complementary technologies in all categories.

Other Applications

A number of small or more speculative applications are included under this heading. They are: (1) the use of coherent plasma oscillations for remote sensing of air pollutants (Lillienfeld, 1977); (2) electrostatic motors; and (3) electrostatic generators.

PATENT SURVEY

A comprehensive survey of modern U.S. patents in the field of corona application has been included in this book (see Chapter 5). The patents are sorted by the following broad subject areas and subareas:

- Equipment/procedures
- Organic synthesis
- Environmental protection
- Surface modification
- Photography, xerography, printing
- Semiconductors
- Textiles, fibers, polymers

Table 1-12. Analytical Competitiveness of a dc Discharge Plasma[a] (Fairless, 1978)

Performance	Plasma	X-ray Fluoresence	Electro-Chemistry	Neutron Activation Analysis	Atomic Absorption
Simultaneous Multielement Analysis	+	L	+	+	–
Ppb PDLS	+	–	+	+	+
Trace-Level Reproducibility, 5%	+	+	+	+	+
Large Linear Dynamic Range	+	+	L	+	–
Assay Suitability	L	+	+	+	–
Matrix Flexibility	+	+	L	+	L
Direct Liquid Organics	+	+	L	+	L
Solids	L	+	L	+	–
Surfaces	–	+	+	–	–
Functional Cost	\$5,000-180,000	\$4,000-50,000	\$1,000 20,000	\$4,000-1,000,000	\$6,000 30,000
Size[b]	0.5-10	0.5-10	0.2-1.5	0.5-1000	1.0
Portability	L	+	+	+	L
Onsite[c]	L	+	L	L	L
Online[d]	L	+	L	+	–

[a]L = Limited capability; + = positive capability; – = no capability.
[b]Size normalized to an atomic absorption spectrometer 45 x 61 x 89 cm.
[c]Small analytical laboratory associated with a production facility. Operator interaction required for analyses.
[d]Automated analyses--no operator interaction.

For each patent we have identified:

- Patent number
- Date of Issue
- Inventor
- Assignee
- Title
- Abstract

CHAPTER 2

VALUES ADDED TO POLYMERIC SURFACES BY LOW-TEMPERATURE PLASMA DISCHARGES

INTRODUCTION

Plasmas can be divided in two basic categories according to their chemical composition:

1. nonpolymer-forming plasmas, usually including elemental gases such as oxygen, hydrogen, helium, argon, nitrogen, air, halogens or gaseous chemical compounds not intended to build polymers; and
2. polymer-forming plasmas, including at least one type of monomer organic compound to be deposited in the form of a polymer.

Following this division there will be two-major areas of plasma utilization for value added to the surface: (1) by chemical reaction of the species in the nonpolymer-forming plasma with the surface; and (2) by deposition of a polymer where the plasma includes polymerizable substances. Both will be examined in the following sections; however, we will limit ourselves to organic polymers when it comes to plasma treatment of surfaces by the action of nonpolymer-forming plasmas.

For the two basic types of reaction mentioned above, plasmas appear to be very similar to various types of radiation, such as gamma rays, ultraviolet light, X-rays, high-energy electrons, which can also create free radicals on polymer surfaces and initiate polymerization of monomers, leading to cross-linking of the polymers exposed, their degradation, and initiation of sites for graft copolymerization of monomers later brought in contact with the surface.

TREATMENT OF ORGANIC POLYMERS BY NONPOLYMER-FORMING PLASMAS

Principles Underlying the Action of Low-Temperature Discharges on Surfaces

We will examine now in some detail the basic types of interaction between nonpolymer-forming plasmas and organic polymeric surfaces. The action of such plasmas on metallic surfaces is rather similar, but is not to be treated in detail here.

The major reactions of the plasma gases with organic polymers can be divided in two main classes.

1. *direct reaction* of activiated gas with polymeric surface (oxidation by oxygen plasma, nitrogen incorporation by nitrogen plasma, halogenation, etc.); and
2. formation of *free chemical radicals* and subsequent reaction of these radicals with the material exposed to the plasma or with another medium which subsequently can be brought in contact with the surface.

The effects may include surface etching, degradation, and graft copolymerization (Yasuda 1976). Combinations of these reactions will also be examined (Bradley 1971). These types of reactions can occur simultaneously in various forms and which one predominates depends on:

1. the chemical composition of the gas;
2. the chemical structure of the polymer;
3. the discharge wattage; and
4. the treatment time.

The new surface properties produced by *direct chemical reactions* depend on the chemically active species. Oxygenation of a surface is usually linked to the following:

- strong adhesive bonds;
- increased wettability by formation of polar groups;
- sites for chemical attachment of other molecules;
- clean removal of polymers; and
- dielectric breakdown if treatment is prolonged.

Nitrogen usually acts as a free radical scavenger and usually is responsible for the destruction of organic compounds and is incorporated in surfaces with the formation of carbon-nitrogen bonds. Fluorine can be incorporated in polyolefinic surfaces and make them nonreactive, like Teflon®.* Hydrogen produces surface reduction and eliminates possible bonds.

---

*Registered trademark of E. I. du Pont de Nemours and Co., Inc., Wilmington, Delaware.

The new surface properties caused by the formation of free radicals are connected with their location. Experiments show that free radicals are formed not only in the surface layer exposed to the plasma, but also at some depth in the bulk material, which demonstrates that electrons, ions and excited gas molecules are not responsible for most of them. In fact, it has been shown that ultraviolet emission from the discharge itself was responsible for most of the free radicals. The free radicals are responsible by subsequent reactions for surface etching, degradation and oxygen incorporation. They are also the centers that initiate graft copolymerization when polymerization monomers are brought into contact with the plasma-treated surface after the discharge. The following is a typical scheme:

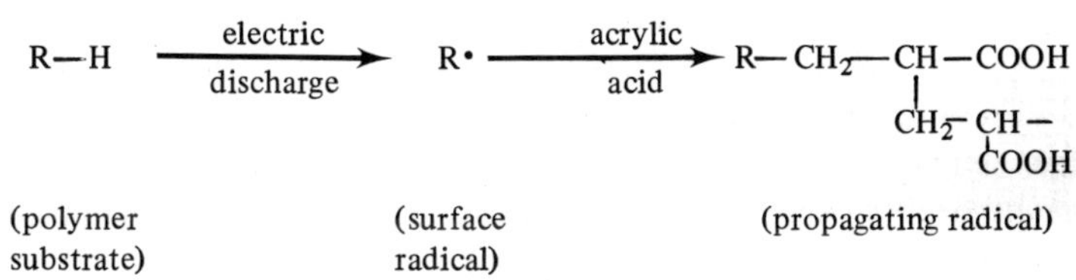

Where the R-H reactor substrate molecule is stripped by the discharge of its H atom, the remaining surface radical is R° and combines with the monomer (here, acrylic acid) to form the propagating radical.

The subsequent reactions to free radical formation all proceed by recombination of these radicals. In the absence of air, the main one is surface cross-linking, building a tough shell. This extends the life of the polymer by its abrasion resistance, renders water-soluble polymers water-resistant, and modifies membrane properties of membranes used as filters. When combined, the two kinds of reactions–direct chemical action and free radical formation–can lead to new properties. As an example, when a polymer is exposed to an argon plasma and then to air, peroxides are formed at reactive sites. Typically, in the formulas below, R-H will represent any organic polymer surface with carbon-hydrogen bonds dissociable in the active plasma:

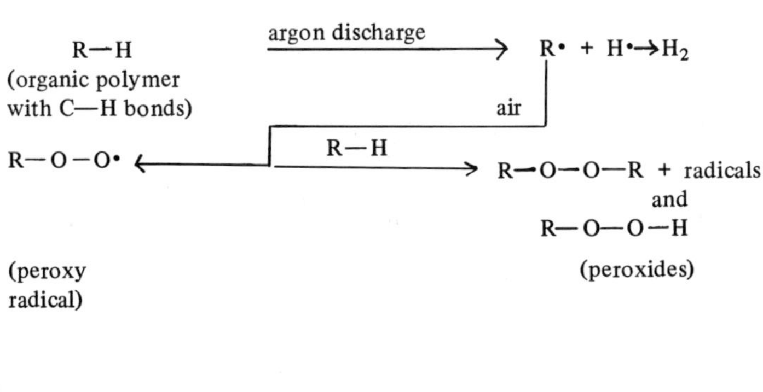

One sees that a peroxy radical is formed after exposure to air, which then reacts with the substrate polymer material to give peroxides. These are known to act as initiators for polymerization and can be utilized to graft subsequently introduced monomers onto the surface by the following scheme, given for the case in which the monomer is acrylic acid:

$$\underset{\text{(peroxide site)}}{R{-}O{-}O{-}H} \xrightarrow[\text{(acrylic acid)}]{CH_2 = CH{-}COOH} \underset{\text{(propagating radical)}}{RO{-}CH_2{-}\dot{C}H{-}COOH} + OH$$

The R-O-O-H peroxide site reacts with the acrylic acid to give the propagating radical and a -OH group. For the grafting reaction to proceed it is sometimes required to raise the temperature, some peroxide groups being indefinitely stable at ambient temperatures. In addition, peroxidized surfaces show improved adhesion and heat sealability, probably because highly reactive oxyradicals are formed by the heat decomposition of the peroxides.

Another example is given by the pressing against each other of two polymeric films treated by plasma. The adhesive properties can be so good that a combination of two polymeric films can be obtained by simple lamination without use of adhesives and remain permanently bonded. Examples are the pairs polyethylene/polyester, poly (ethylene vinyl acetate)/polyester. Pouches can be prepared in this manner and used for boil-in-bag frozen food packages.

Nearly all polymers lose weight when exposed to plasmas, depending on the type of gas used. The loss is proportionate to the time of exposure and strongly dependent on the discharge wattage, probably because of the role of the UV emission that is enhanced by higher power in the discharge.

A review of the behavior of nonpolymer-forming plasmas is presented in Table 2-1. Schematic representation of the interaction of a nonpolymer-forming plasma with a polymer substrate is presented in Figure 2-1.

Applications of Nonpolymer-Forming Plasma Treatments with Organic Polymers

The following modifications of surface properties have been found to be possible depending on the gaseous composition, pressure, duration and voltage of the discharge, and the nature of the surface material:

- Wettability
- Molecular weight variation
- Chemical composition change
- Improvement of adhesive properties

Table 2-1. Behavior of Nonpolymer-Forming Plasmas

Type of Action and Process	Effects	Leading Chemical Reaction	Practical Consequences
Direct action of activated plasma gas on polymer.	Surface oxidation.	Oxidation.	Strong adhesive bonds. Sites for chemical attachment of other molecules. Clean removal of polymers.
	Nitrogen incorporation.	Active nitrogen acting as radical scavenger.	Destruction of organic compounds. Incorporation of nitrogen in surface to form carbon-nitrogen bonds.
	Halogenation, reduction.	Fluorination.	Production of chemically nonreactive surface.
Free radicals formation in polymer bulk by UV irradiation and subsequent reaction.	Surface etching, degradation, oxygen incorporation, graft copolymerization.	Recombination of free radicals. Surface cross-linking in absence of air.	Limits plasticizer loss and extends life of plasticized products. Renders water-soluble polymers water resistant. Modifies membrane properties. Initiates grafting (copolymerization).
(1+2 + air). Peroxide formation (first exposed to argon plasma, then to air).	R-O-O sites react with original polymers to form peroxides.	Peroxide acts as initiator. Peroxide thermally decomposes to form highly reactive oxyradicals.	Initiates grafting (copolymerization).
(1+2 + surfaces). Interfacial recombination: pressing of 2 previously plasma-treated surfaces against each other by lamination.	Permanent bonding.	Many of the above, in particular reactive sites.	Improves adhesion and bonding. Heat sealability. Enhanced bonding through reactive sites.
(1+2 + monomer). Grafting (copolymerization by exposure of plasma-treated polymer to monomer.	Copolymerization of new polymers with original polymer substrate.	Free radical and peroxide sites act as initiators.	New polymer application on original polymer substrate.

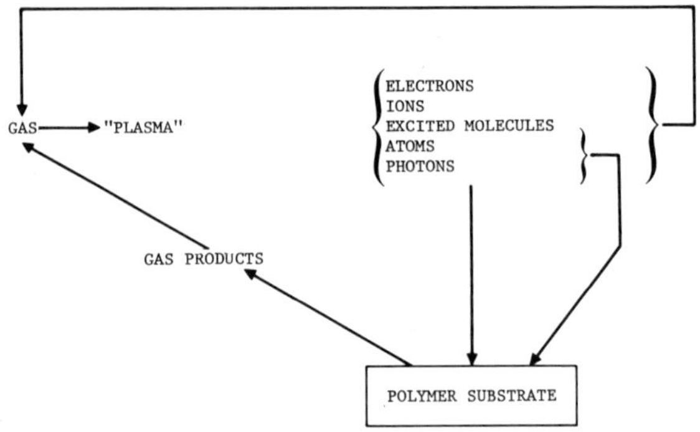

Figure 2-1. Interaction of plasma with a polymer surface.

- Microroughening
- Shrinkproofing of wool and hair
- Cleaning
- Disinfection
- Porosity
- Layer removal (plasma ashing, microincineration)
- Layer etching
- Grafting of monomers (copolymerization)

Wettability of Polymer Surfaces

The wettability of a surface relates to the adherence of inks, adhesives and dyes. It is considered that changes in wettability of polymer surfaces are caused by attaching functional groups to the surface. Oxygen-containing groups are found to be more active than nonoxygen-containing ones. These functional groups have surface densities comparable to the monomer density. Plasma action is relatively fast (seconds) compared to the thermal process, which requires minutes to hours for the same change in wettability. Antioxidants usually placed in commercial polymers do not affect plasma oxidation, unlike thermal oxidation.

The attachment of groups not containing oxygen can be achieved in two ways. One is to expose the polymer to a nonoxygen-containing plasma. Another is to expose the polymer to air-inert gas discharge to prepare the surface, and then to dip it into a solution containing a component capable of reacting with the polymer-free radicals.

With regard to surface phenomena associated with printing and bonding, all solid matter is classified as having high or low surface energy. A drop of water deposited on a clean metal or glass surface will spread and undergo a marked increase in liquid surface area. The same drop deposited on a wax or untreated plastic surface will retain most of its original spherical shape. The metal or glass surface could perform sufficient physical work on the water to completely overcome liquid surface tension forces. Such a solid surface is classified as a high energy type. The wax or plastic surface could not perform as much physical work on the water and is classified as a low-energy type (Zisman, 1964).

Figure 2-2 illustrates a water drop resting at thermodynamic equilibrium on both high- and low-energy surfaces. Also noted are other terminology used to describe such surfaces. The terms wettable-unwettable and hydrophilic-hydrophobic serve to only describe our observation of a static state. The energy or thermodynamic concept is more in keeping with scientific methods of thinking. On high-energy surfaces the contact angle is small; on low-energy surfaces it is large.

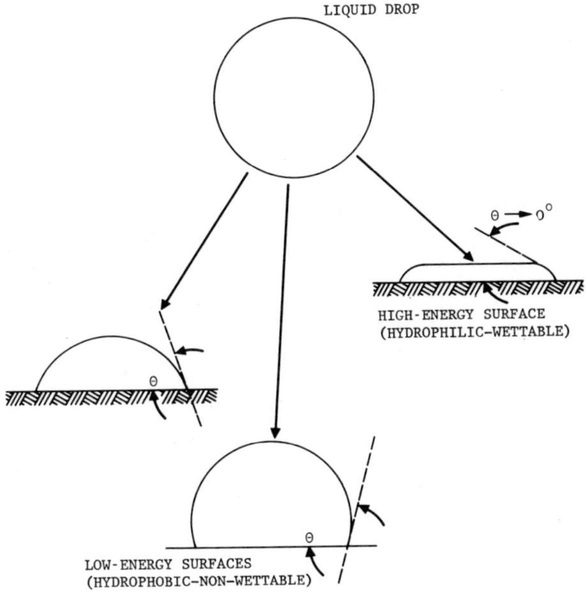

Figure 2-2. Water drop in thermodynamic equilibrium.

The atoms of hard materials, such as metals, glass or ceramics, are bonded by strong primary valence forces (metallic and ionic). The secondary valence forces extending beyond thin physical confines are likewise strong and provide high surface energies. By comparison, the molecules of natural and synthetic organic polymers (softer materials such as wood, rubber, wax, plastic, etc.) are bonded by weaker primary valence forces (the convalent bonds of carbon with itself and with oxygen, hydrogen, etc.). The resultant secondary valence forces are weaker and provide lower surface energies.

Objects with high surface energies are more readily printable; the others require surface treatment to render them printable. A list of plastic films in decreasing order of surface energy is provided in Table 2-2 with an accompaning representation of molecular structure (McLaughlin, 1966).

Table 2-2. Some Plastic Films and Their Surface Energy (McLaughlin, 1966)

Name	Chemical Formulation	Surface Energy (dynes/cm)
Nylon (polyamide)	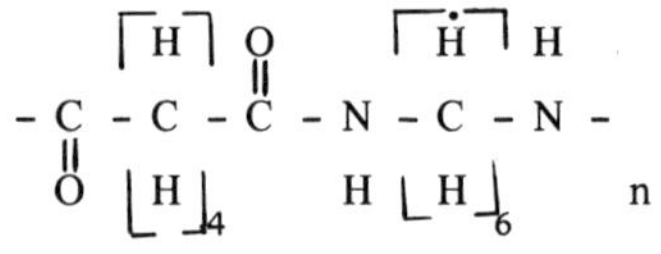	42
Polyester (Mylar®)	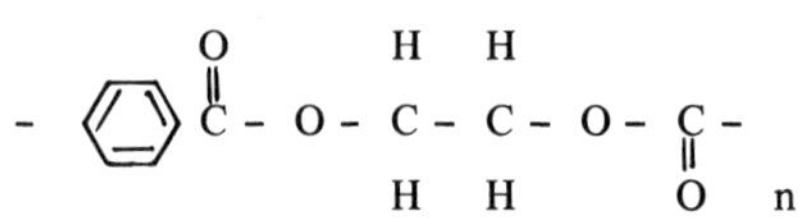	41
PVC (vinyl)	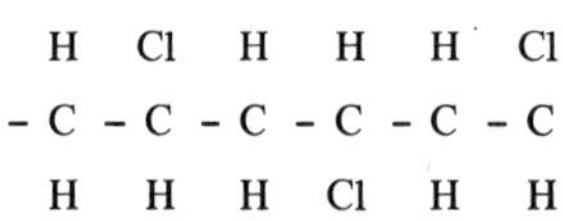	38
Ionomer (Surlyn A®)	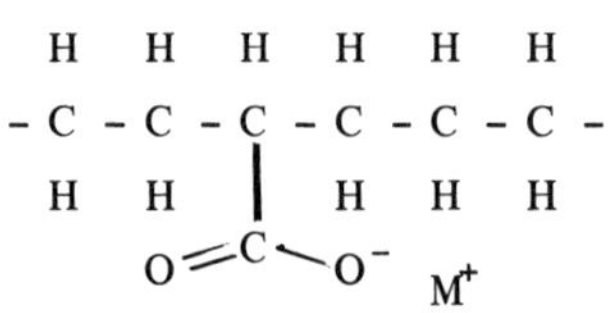	38
Polystyrene	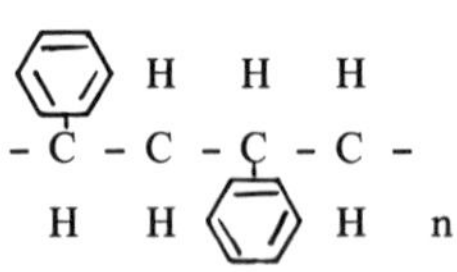	33
Polyetheylene	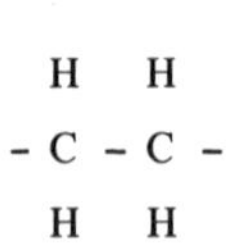	31
Polypropylene	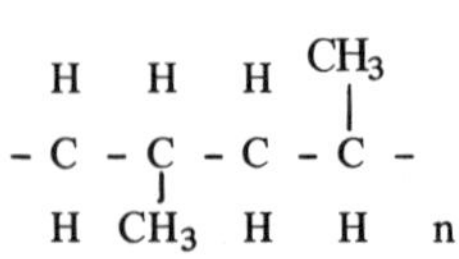	29

Plastic materials such as nylon and Mylar® having higher surface energies have more pronounced dipole moments in their structure. The polyolefins and polystyrene are quite noticeable at the lower end of Table XIV, while copolymers and the new ionomers are just above them. These materials have lower surface energies and require surface treatment. The straight hydrocarbon (-CH_2-) structure of polyolefins has no strong dipole moments, weaker secondary valence forces and lower surface energy. To give printability to low-energy surfaces, the molecular structure of the surface microlayer must be altered to raise the magnitude of the surface energy.

With thermoplastic articles being formed by the removal of heat from a molten mass, their surface cools faster than the interior and has a tendency to accumulate the lower molecular weight components of the polymer as well as a higher proportion of any specific additives. The intense heat generated by the electrical discharge serves to "flash off" many of these low-molecular-weight components and additives and thus cleans the surface (Bickerman, 1950).

Chemical changes induced by treatment are primarily polymer chain scissions, formation of polar groups (carbonyl, aldehyde, hydroxy) and unsaturation (Hines, 1957; Rosman, 1956). Treatment takes place at certain existing sites on the surface. When these are consumed, additional exposure to the electrical discharge makes no improvement, but rather hardens the surface (McKelvey, 1962).

Those polymers that have been studied for changes in wettability (and also in molecular weight and surface chemistry) with various plasma gases are shown in Table 2-3.

Plasma processes leading to improvement in *surface wettability,* i.e., in the adherence of dyes, inks and adhesives to surfaces, have been applied in the packaging, electronics, construction and clothing industries. Polymer surfaces tend to be inert and cannot be painted, dyed or printed. Labeling of polymer items is a major problem, particularly of jackets of wires and cables. Ink color will only adhere after plasma treatment of the surface, which has the advantage of producing no change in the bulk properties of the material, something that would be altered by a thermal treatment.

The following is an example of a plasma process for treating plastic films and coated cellulosic web (McLaughlin, 1966). The moving web is passed over a grounded metal roll, which has on its cylindrical surface a continuous, solid insulating cover. Above the cover and parallel to its axis a long metal electrode is mounted. The lower, flat surface of the electrode establishes a uniform air gap with the exposed surface of the film (Figure 2-3). The treating process is normally performed in line with the windup after the extrusion or coating of the web.

From an electric viewpoint, the metal electrode surface and that segment of the metal roll (both conductors) facing it are the plates of a capacitor

Table 2-3. Polymers Studied with Various Plasma Gases (Hollahan and Bell, 1974).

Polymer Material	Common Name	Wettability						Molecular Weight Changes						Chemical Reactivity
		Oxygen	Hydrogen	Nitrogen	Helium	Argon	Inert Gases	Oxygen	Hydrogen	Nitrogen	Helium	Argon	Inert Gases	Reagent
Fluorocarbon Polymers														
Polytetrafluorethylene	TFE, Teflon	x	x	x	x	x	x	x	x	x	x	x	x	NH_3
Poly (vinyl fluoride)	Tedlar	x						x	x		x			
Polyperfluoropropylene	Pep, Teflon	x						x			x			
Poly (vinylidene fluoride)	Kynar	x						x			x			
Polytrifluorochlorethylene	Kel-f	x						x						
Hydrocarbon Polymers														
Polyethylene		x	x	x	x	x	x	x	x	x	x	x	x	CO_2
Polypropylene		x	x	x	x	x	x	x	x	x	x	x	x	NH_3, N_2O, H_2S
Polystyrene								x			x			
Polyvinylcyclohexane								x						
Polyisobutylene														
Polybutene-1										x				
Polybutadiene		x						x						
Hydrocarbons (oxygen, nitrogen)														
Poly (ethylene terephhtalate)	Mylar[a]	x						x		x				
Polycarbonate	Lexan[b]	x						x		x				NH_3
Poly (oxymethylene)	Delrin[c], Celcon[c]	x						x		x				
Cellulose Acetate								x						
Poly (methy methacrylate)	Plexiglas[d]													

Polyhexamethylamine	Lucite[b]					x		NH_3
Polyurethane	Nylon[e]	x				x	x	NH_3
Hydrocarbon (chlorine)								
Poly (vinylchloride)								NH_3
Hydrocarbon (silicone, oxygen)								
Polydimethylsiloxane		x	x	x	x			NH_3

[a] E. I. du Pont de Nemours and Co.
[b] General Electric Co.
[c] Celanese Plastic Co.
[d] Rohm and Haas Co.
[e] Resin Name

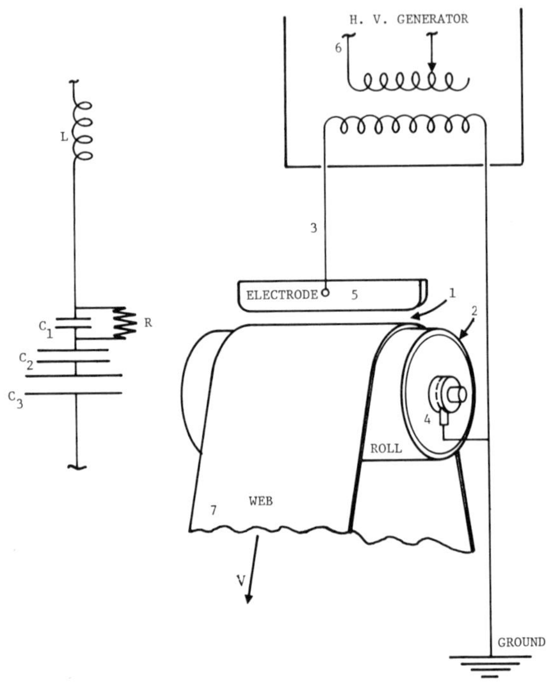

Figure 2-3. Plasma treatment of continuously moving web.

(condenser). The air gap, plastic web and roll cover comprise dielectric (nonconductive) layers between the parallel capacitor plates. Where a source of alternating voltage is applied between electrode and ground, an alternating current is sustained in the capacitor circuit because of the bound charges in the dielectrics. To induce treatment, a sufficient voltage (usually greater than 5000 volts) is applied between electrode and ground, so as to exceed the dielectric strength of the air but not that of the web or roll cover. The exposed surface of the web is thus given an exposure to the ionization and heat of the electric breakdown of the air as it passes under the electrode. The best operation of the electrical discharge treating system depends on using the correct equipment and installation.

Engineering experience has shown that all of the following factors are important in treating plastic films (McLaughlin, 1966):

1. Maintain minimum air gap between electrode surface and that of web. Recommended setting is 0.063 in. (1.6 mm), but never exceed 0.130 in. (3.3 mm) if possible.
2. Use a roll cover of 0.190 in (3.3-4.8 mm) compounded "Hypalon" chlorosulfonated polyethylene elastomer. The dielectric constant

of this material should have minimum values of 8 at a frequency of 100 cycles/sec and 6 at a frequency of 1 megacycle.

3. Install a high-voltage generator close to the electrode to permit minimum lead wire. All leads must be supported by high-voltage insulators, so as not to touch metal structural components.
4. Provide a direct ground connection body of metal roll by contact ring and brush.
5. For web speeds in excess of 150 ft/min (46 m/min), multiple electrodes are required.
6. Power requirements for treatment vary, depending on the material and its processing history. Polpropylene and polyethylene film require about 1 $w/ft^2/min$ (10.8 $w/m^2/min$). However, if these same materials have been extrusion-coated onto a substrate web, their surface thus experiencing high temperatures, treating requirements could be as low as 0.5 $w/ft^2/min$ (5.4 $w/m^2/min$). The higher vinyl acetate copolymers of ethylene and new ionomers, such as DuPont's Surlyn® A-ionomer resin require less treatment than equipment films or coatings of straight hydrocarbon polymers.
7. Once treatment is imparted to any plastic web, any abrasion of the treated surface with subsequent handling should be avoided. On windup, the tension should be the minimum to control roll formation.

A graphic plot of contact angle vs solid surface tension is presented in Figure 2-4. Imposed on this plot are diagonal lines representing the various plastics previously presented in Table XIV. Also imposed are vertical lines representing water (72 dynes/cm) and glycerol (63 dynes/cm). For this plot, two factors can be pointed out, both of which can be utilized in measuring the treatment of plastic surfaces by wettability:

1. Using a control liquid such as water, the contact angle is an inverse measure of surface properties. Plastics with higher solid surface tensions define lower contact angles.
2. The intercept of each diagonal line with the upper horizontal line representing zero contact angle is, by definition, the critical wetting tension of that plastic surface. Liquids of varying surface tensions can thus be used to measure solid surface properties.

When polyethylene film is surface treated a series of new diagonal lines represent each level of treatment intensity. A duplication of Figure 2-4 is presented in Figure 2-5. Also included are diagonal lines representing the treatment of an ionomer film (Surlyn A-1601). The new graph shows that contact angles of 96, 94, 90, 80 and 60° are measurements of surface treatment for polyethylene, while 83° and a 53-49° range are measurements for the ionomer film. This graph also shows that liquids of the following surface tensions could be used to measure the treatment levels on the polyethylene film; 31, 34, 37 39, and 42 dynes/cm. The ionomer film could likewise be measured with liquids of 38, 49 and 54 dynes/cm. This procedure is called the critical wetting tension test or Visking Test.

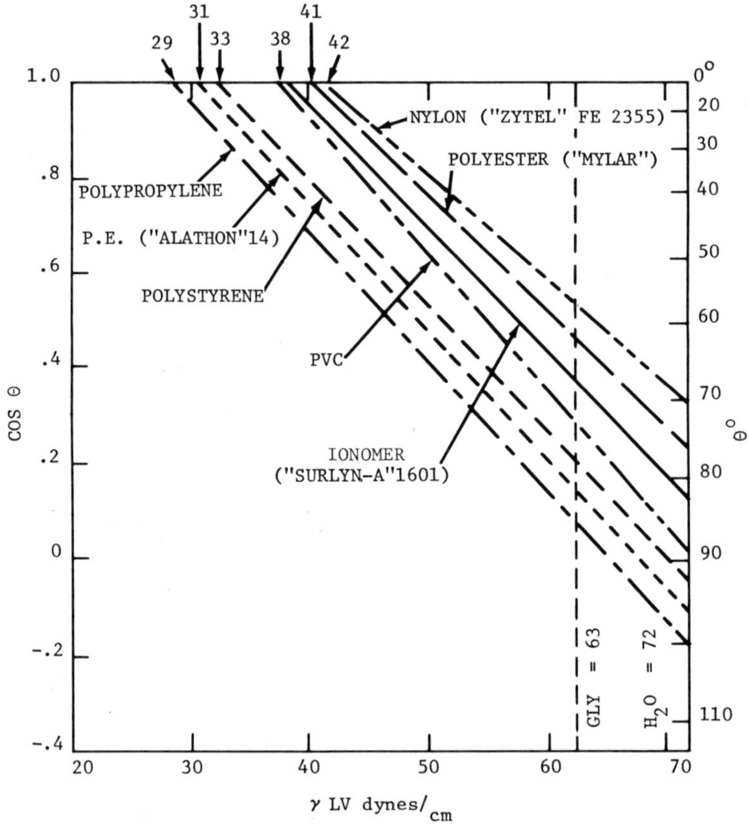

Figure 2-4. Contact angle vs solid surface tension for various polymers (McLaughlin, 1966).

The organic surface materials to which plasma treatments have been applied include:

- Polymers
- Natural materials (textiles)
- Celluslosic materials (papers, films).

Surface Layer Molecular Weight Variation

The molecular weight of a polymer placed in a plasma increases for noble gas, nitrogen and hydrogen plasmas due to cross-linking, while it decreases for an oxygen plasma due to scissioning.

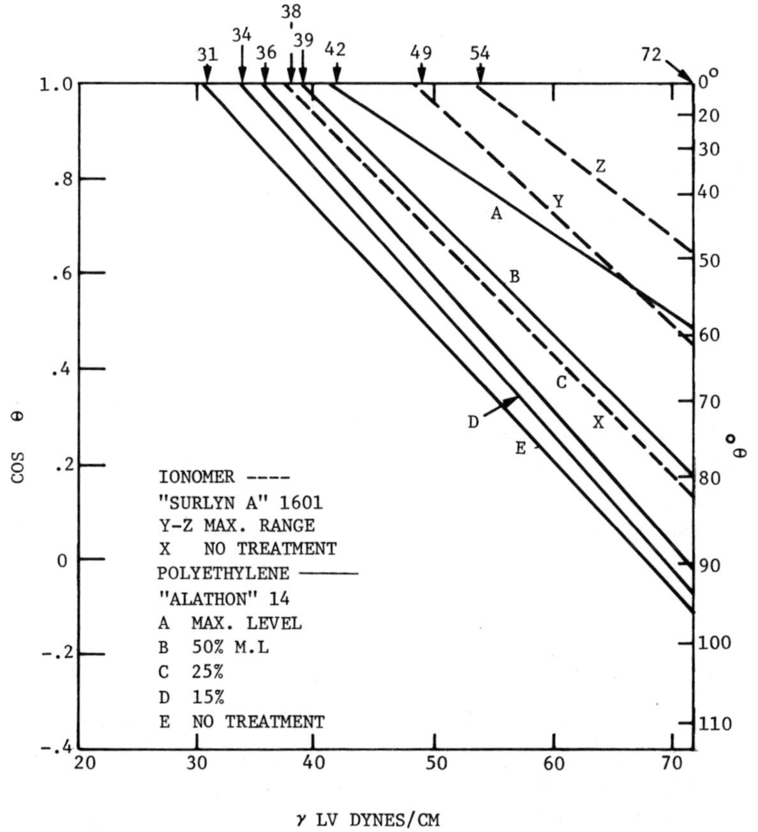

Figure 2-5. Improvement of solid surface tension after plasma treatment (McLaughlin, 1966).

For nonoxygen-containing plasmas, the depth within which molecular weight increase takes place is 1-10 microns. Cross-linking in the layer has been demonstrated to be due to the ultraviolet radiation produced in the plasma and not to gas atoms. The effective wavelengths are less than 1900 Å. Surface properties such as wettability are unchanged when inert gases are used.

Oxygen plasmas produce free radicals that can cross-link but do not rebond (branch, cross-link, recombine) because of subsequent oxidations. Cross-linking is, however, produced by UV in the region 500 Å thick where the UV intensity is large (near the surface).

Oxygen atoms readily diffuse 1-10 microns into the polymer below this layer and tie up all the free radicals, producing a decrease in molecular weight. In addition, there is production of volatile compounds that ablate from the surface, producing a continuous loss of material linear with time. This appears to be an exclusive property of oxygen-containing plasmas. Unsaturation can be produced simultaneously with molecular weight changes. Higher polymer temperature increases the degree of cross-linking. Changes in molecular weight depend on the number of bond changes per polymer molecule, which is a function of the average molecular weight.

Variations in molecular weight of a layer at the surface of the polymer generally affect physical properties such as permeability, solubility, melt temperature and cohesive strength (which is important for adhesion). Because of improved adhesion, bonding of polymers with other materials is also made possible by plasmas. Silicon as used in electronics has a weak environmental resistance and can only be sealed in epoxy resins after it has been exposed to plasmas. Metallic electrodes, such as silver, can only be satisfactorily vapor-deposited on polymers after the polymers have been corona-treated. Other examples are silicone and Teflon bonded to polyethylene, and polymers bonded to epoxies, which, in turn, can be bonded to a large number of other materials.

Surface molecular weight changes due to cross-linking of a surface layer by UV radiation are achieved on a large industrial scale for polyethylene objects.

Changes in Surface Chemistry

Nonoxidative changes in the surface composition can be traced to:

1. covalent bonding of a chemically active group; or
2. changes in molecular structure, consisting in rearrangements leading to unsaturated sites.

As examples of covalent bonding, fluorination was found to give a non-reactive surface like that of Teflon. The carbonyl (-C=0) group was also successfully added to the surface. Attachment of amines and halogens by covalent bonding to the surface acts as "hooks" for the addition of new compounds. As an example, capillary tubes of inert polymers cause blood clotting. Anticoagulants like heparin can be added to the surface if amine groups have been attached to it by corona discharges. Deposition or "grafting" of polymer films onto the surface also depends on changes of surface chemical composition. For this application, one can either introduce the gas or vapor of the monomer to grafted into the discharge or only prepare the substrate with a discharge to form free radicals, which would thn initiate polymerization when the monomers would be brought in contact with the surface.

It has been demonstrated that unsaturation occurs to the same depth as cross-linking, which shows that UV radiation is an important participant in the unsaturation mechanism.

Porosity Increase

Large porosity is a prerequisite to bacterial growth and makes the polymer compatible with living tissue. Porosity can be adequately controlled by the action of low-temperature gas discharges (Yasuda 1976, 1977). Applications include preparation of culture trays for bacteria, medical supplies, in particular objects to be implanted in the body, and filter membranes for electrodialysis, reverse osmosis and ultrafiltration.

Plasma Treatment of Natural Materials

A whole area of applications of plasma discharges occurs with natural, mostly cellulosic materials. Many natural polymers such as wool, mohair (goat wool), leather, cotton (Abbott and Robinson, 1977) and paper have met increasing competition from synthetic materials. A portion of this competition was offset by surface treatments of the natural polymer in plasmas to give them properties achieved by the synthetic materials.

The effects of electric discharges do not penetrate below 100 Å from the surface in synthetic polymers and probably also not in natural ones.

Microroughening by glow discharges can eliminate too smooth surfaces of natural fibers. Corona discharges have been applied with success to cellulosic textile fibers. They eliminate the normal wax coating, roughen the fiber surface, and increase the breaking strength from 980 g to 1280 g for optimal twist (Stone and Barrett, 1962).

Adhesion (bond-strength) of coatings on fibers and cellulosic flat surfaces can be improved. Goring (1967) found that the wet-bonding strength of cellulose strips increased from 11 to 80 kg/cm^2 in 5-10 minutes at 50° C. A little moisture seems necessary since the same effect is not found at 105° C. The heat-bonding strength of cellulose to polymers can be increased 5-7 times by corona treatment in air and 15 times in nitrogen (Kim et al., 1970).

Corona discharges can be used for preparing paper at 6-40 ft/sec to receive a polyethylene coating, which enjoys excellent adhesion even when applied 24 hours later (Green, 1965).

The *friction* coefficient of cotton fibers was found to increase 28-32% when treated with an air-chlorine corona (Thorsen, 1971). Adhesion between fibers was increased 323% for 5 minutes, then decreased to 150% relative to untreated fibers. Thorsen also investigated wool and mohair (goat's hair) with the same air-chlorine corona and found a 230% increase in cohesive forces between fiber for wool, which decreased to 38%, and a 95% increase for mohair, which remained constant. All these fibers were loosely woven, since corona discharge penetration into yarn or woven fabric is limited.

Natural fibers have an outer shell (cuticle) working as a hydrophobic layer. If it is modified or removed, water absorption capability is increased, while rate of dyeing and soil repellency are improved. When plasma treatment is allowed to modify the epicuticle of natural fibers, the electropositive wool surface can be made more negative, which improves the soil repellency of wool and mohair. The surface becomes wettable, hydrophilic and displays increased dyeability. Brown and Swanson (1971) have scientifically demonstrated that the surface-free energy of corona-treated cellulose is significantly increased due to an increase in the polarity of the surface, which was observed to be more hydrophilic after application of the electric discharge.

Felting Shrinkage of wool and mohair is related to frictional and hydrophilic characteristics of the fibers. Making the fiber smoother and more hydrophilic reduces shrinkage. Felting shrinkage occurs when wool is washed while being agitated, and this imposes limits on its use. The individual fibers move with respect to each other and become entangled. Therefore, means to limit the extent of shrinkage have been sought. A commercially acceptable shrinkproofing process was developed by the U.S. Department of Agriculture (Pavlath et al., 1971). Radiofrequencies of 13.56 MHz were used. Area shrinkage on the order of 10% was obtained for products of power and residence time in the range 10-30 watt seconds. The nature of the gas in which the discharge takes place does not seem to be relevant, which implies that the active species are the electrons rather than the ions.

Grafting of polymers on natural materials is made possible by plasmas. The best approach seems to expose only the substrate to the glow discharge and to contact later the treated surface with the monomer. Akutin et al. (1960) have reported grafting of acrylates to cellulose in a corona discharge. This method lends itself to grafting volatile monomers. Simionescu (1969) grafted polyacrylonitrile to cotton in a glow discharge in argon at 45 MHz. Byrne et al. (1972) grafted cotton with di- and tetrafluoroethylene to obtain a water-repellent surface.

Cleaning, Disinfection, Ashing, Etching

When it comes to stains, impurities or deposits on substrates, gas discharges are an efficient way of removing them without damaging the substrate. They also sterilize. Highly branched saturated hydrocarbon polymers, which do not react with molecular oxygen below 600° C, react with the atomic oxygen produced in the discharge at 40° C and can be removed (ashing).

It was also found that the degradation proceeds in a very gentle fashion, without sputtering or mechanical damage to delicate inorganic structures, even on a microscopic basis, which makes plasmas effective for use in sampling materials for chemical and physical analysis and microscopy.

The plasmas used are either oxidizing (pure oxygen) or reducing (pure hydrogen). Application of oxidizing plasmas in analysis include:

- dry-ashing of organic, biological and polymeric material;
- removal of organic materials bearing the organic residue for subsequent analysis;
- reaction of samples with plasma to produce gaseous products for subsequent analysis;
- preconcentration of organic materials to reduce their volume before subsequent analysis; and
- cleaning of surfaces for regeneration of their contaminant-free condition.

Hydrogen plasmas can also be used to degrade carbonaceous hydrocarbons or protein products, although at a lower rate than oxidizing plasmas. The advantage is total avoidance of oxidation of the sample.

Binding of Plastic to Metal

Usually this is done by pressing plastic sheet stock to metal with an intermediate adhesive. The difficulty here is that the adhesive used between the metal and the polymer in most cases does not get a good bonding to the polymer. Low-temperature plasma treatment of the polymer sheet ensures good bonding of the adhesive to the polymer surface, due to the creation on the polymer surface of free radical sites, which facilitate chemical grafting of the adhesive. With plasma-treated polymer surfaces adhesion values obtained were outstanding. Hall et al. (1969) used helium, oxygen and nitrogen plasmas to treat the surface of various polymers before bonding it to aluminum and obtained up to the following average values for lap shear bond strengths (in psi) depending on treatment gas:

High-density polyethylene	3770 (He), 3165 (O), 2510 (Cl), 3500 (N)
Nylon 5	3960 (He), 3490 (O)
Celcon	1220 (He), 1260 (O)
Cellophane	1430 (He), 1400 (O)
FEP Teflon	1730 (He), 2260 (O), 1080 (N)
Polyprophylene	2600 (He), 3080 (O)

There are numerous applications of plastic-covered metal sheets in the industrial area. Once the plastic sheet has been bonded to the metal sheet, the composite sheet can be worked into forms and shapes and used to make flasks, boxes, containers and parts. In the cosmetic industry, containers made of plastic-covered metal tend to be used for all alcoholic perfume solutions instead of lacquer-coated metal containers. The alcoholic solvent usually attacks the lacquer coating, and the solution is then free to stain the container surface. When the lacquer is replaced by a plastic coating sheet, which is not subject to attack by alcohol, the container's useful life increases dramatically. Medical applications are also numerous. In the electrical connector industry, Teflon-to-metal bonds are paramount. The automotive, aircraft, decorative and jewelry industries also use great amounts of plastic-covered metal stock.

Plastic coatings on metals are also useful where salt or other corrosion hazards exist, such as in marine applications, in the chemical industry (where corroding vapors and fumes are frequent), and in all high-technology areas (other than the semiconductor industry where low-temperature plasmas have found quite general use, although most of it is not for polymer on metal coating). In the food industry, many metal cans have recently appeared with plastic coating on the interior; they are made from plastic-coated metal sheet. By contrast, Teflon-coated frying pans are made by dipping.

Another method to obtain intimate bonding of polymer and metal is to directly deposit the polymer on the metal with a plasma discharge where the monomer has been introduced as a reactant gas. As an example, some headlights for automobiles consist of aluminum, on which a very resistant polymer coating has been plasma deposited. A large variety of monomers for coatings can be polymerized and have been investigated. Extensive work is underway in research laboratories, both of the manufacturing industry and the government (especially NASA Ames laboratory) to develop the best plasma-deposited coatings; however, this work is currently proprietary. There are, as yet, very few direct industrial applications of polymer deposition by plasmas, because the technique has not been perfected.

PLASMA POLYMERIZATION

Principles Underlying the Deposition of Plasma Polymers

When an organic vapor is introduced into a plasma (argon and helium are mostly used in these cases to substain the discharge) or if a plasma of the organic vapor can be produced without inorganic gas addition, polymerization of the organic compound occurs in the plasma and the polymer-formed deposits. It can deposit on the walls of the vessel or on a substrate of another polymer, in which case it is highly cross-linked and bonded to the substrate, providing a good surface coating or grafting into the substrate polymer.

The mechanism of this polymerization in a plasma is quite different from conventional monomer polymerization and also depends on plasma conditions (temperature, geometry). Basically, the three following routes seem to be present in various proportions in all cases (Yasuda, 1976):

1. The monomer undergoes fragmentation in the plasma before or during the polymerization. As an example, acrylic acid loses carboxylic acid groups.

2. Polymerization occurs by a unique mechanism called plasma polymerization in the plasma.
3. In some cases (vinyl monomers), the conventional addition mechanism could take place, initiated by the active species in the plasma (plasma-induced polymerization, which can proceed after extinction of the plasma if monomer is present so long as there are reactive sites with free radicals still available for recombination).

Table 2-4 gives deposition rates of vinyl monomers and their saturated compounds; the differences in rates of the former (monomer) are due to plasma-induced polymerization, which does not occur with the latter (saturated compounds). As these differences are not great by and large, the genuine plasma polymerization is the major mechanism rather than the plasma-induced conventional polymerization (by an addition scheme).

Table 2-4. Comparison of Polymer Deposition Rates[a] for Vinyl and Saturated Vinyl Compounds (Yasuda, 1976)[b]

Vinyl Monomer		Saturated Vinyl Monomer	
Compound	$k \times 10^4$ (cm^{-2})	Compound	$k \times 10^4$ (cm^{-2})
4-Vinylpyridine	7.59	4-Ethylpyridine	4.72
α-Methylstyrene	5.33	Cumene	4.05
Styrene	5.65	Ethylbenzene	4.52
N-Vinylpyrrolidone	7.75	N-Ethylpyrrolidone	3.76
Acrylonitrile	5.71	Propionitrile	4.49
Vinylidene Chloride	5.47	1,1′-Dichloroethane	2.98
Allylamine	2.86	n-Butylamine	2.52
Methyl Acrylate	0.99	Methyl Propionate	0.57

[a]Polymer deposition rate, R, in g/cm^2 min is given by $R = kF_W$; F_W is the weight basis flowrate (g/min) 4, 5.
[b]Reprinted by courtesy of Marcel Dekker, Inc.

Plasma polymerization is supposed to proceed mainly by the coupling of primary radicals or excited species generated by the ionization of monomer vapor. According to this concept, the following two steps can summarize the phenomenon:

1. Initiation:

$$M_i \longrightarrow M_i^{\circ}$$
$$M_k \longrightarrow M_k^{\circ}$$

2. Propagation:

$$M_i^{\circ} + M_k^{\circ} \longrightarrow M_i - M_k$$
$$M_i^{\circ} + M_k \longrightarrow M_i - M_k$$

where i and k are numbers of repeating units, i.e., i=k=1 for the monomer, and the M° are reactive species, which can be ions of either charge, excited molecules or free radicals. These are stepwise reactions (no chain reactions). The product of a reaction can be subjected to repeated initiation processes (Yasuda, 1977; Tibbitt et al., 1977; Lam et al., 1976; Hudis, 1957).

The free radicals to be formed before polymerization occurs, when they become coupled, can be produced in three ways:

1. opening of a double (or triple) bond;
2. hydrogen abstraction (detachment of a hydrogen atom from a monomer); and
3. cleavage of a C-C bond in a cyclic compound (cleavage of such a bond in a noncyclic compound will contribute to the formation of a polymer).

Pressure changes in the closed system are a convenient way to investigate the contributions of these three paths:

1. Opening of a double or triple bond leads to a decrease in pressure and no hydrogen production because it does not change the total number of molecules.
2. Detachment of hydrogen atoms leads to no change in pressure because production of hydrogen gas molecules compensates for loss of an inorganic molecule by polymerization.
3. Cleavage of a C-C bond in a cyclic compound leads to a decrease in pressure and no hydrogen production just as opening of a double or triple bond.

This is not only found valid for hydrocarbons, but also for most nitrogen-containing organic compounds, such as amines and nitriles.

Researchers have thus been led to investigate the hydrogen yield, i.e., the number of hydrogen gas molecules liberated from a molecule of monomer by the plasma polymerization process. (In the conventional addition process no hydrogen is liberated.) Figure 2-6 shows the hydrogen yield as a function of the number of hydrogen per multiple bond and/or cyclic structure in the monomer. Obviously, there are chemical structure groups of monomers behaving differently according to the following three curves obtained:

1. Aromatic, heteroaromatic and triple bond-containing structures lie on the lowest yield curve.
2. Compounds with cyclic structure and/or double bond lie on the intermediate yield curve.
3. Saturated compounds lie on the highest yield curve.

By comparing the three curves and the monomer they include, one arrives at the following general conclusions (Yasuda, 1976):

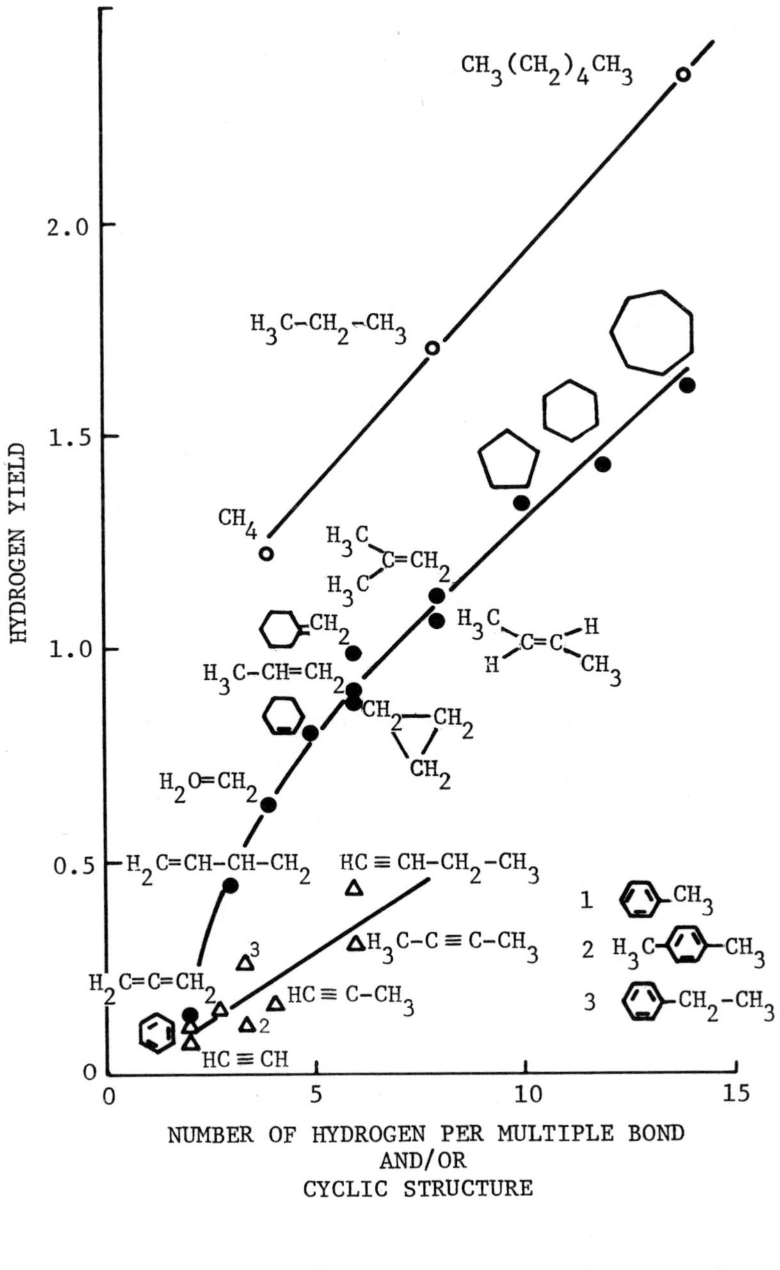

Figure 2-6. Hydrogen yield observed with the plasma polymerization of organic compounds (Yasuda, 1976). Reprinted by courtesy of Marcel Dekker, Inc.

1. The more striking fact is that compounds with triple bonds or conjugated double bonds polymerize with a very small hydrogen yield. This means that the polymerization of aromatic compounds must proceed by the opening of double bonds in the benzene ring.
2. Compounds with olefinic double bonds show a relatively high hydrogen yield, indicating that the contribution of polymerization by the opening of double bonds is surprisingly small. In the cases of ethylene and propylene, the role of hydrogen abstraction is almost equal that of the double-bond opening.
3. The hydrogen yield of cyclic compounds is smaller than that of corresponding normal compounds. The contribution of the opening of a cyclic structure to polymerization is similar to that of an olefinic double bond, as in Figure 17. (Reprinted from Yasuda, 1976, courtesy of Marcel Dekker, Inc.)

The mechanism of genuine plasma polymerization implies that many small molecules (N_2, CO_2, H_2O) can be copolymerized or included into the plasma polymers. Table 2-5 shows some experimental results to confirm this.

Free radicals are still found in plasma polymers after plasma polymerization has occurred. Their presence is explained by the difference between production of free radicals in the discharge and consumption of free radicals by the recombination mechanism. A high concentration of free radicals is found in polymers from compounds on the lower hydrogen yield curve, fewer from compounds on the intermediate curve and very few from compounds on the higher yield curve.

A schematic representation of the interaction of polymer-forming plasma with a polymer surface is shown in Figure 2-7 (Yasuda, 1976).

Advantages, Properties and Applications of Thin Film Deposition by Plasma Polymerization

An advantage of plasma polymerization is the wide variety of compounds available for film deposition through this technique. Bradley and Hammes (1963) concluded that, in order to be plasma polymerizable, a monomer should contain either a carbon-hydrogen bond, a carbon-carbon double bond or an aromatic ring, unless it was perhalogenated. A qualitative rating of a variety of monomers for producing polymer films is given in Table 2-6 (Smolinsky et al., 1968). Investigations show that if the monomer feed is gradually varied in chemical composition, the composition of the intervening film was varying uniformly in the same proportions. The effect of diluting the monomer with a carrier inert gas such as argon is that higher concentrations of free radicals are formed, resulting in increased cross-linking of the films. Small additions of halocarbons result in increased polymer yields. Polymerization efficiencies of monomers are given in Table 2-7 (Bradley and Hammes, 1963).

Table 2-5. Results of Elemental Analysis of Some Plasma Polymers (Yasuda, 1976)[a]

Plasma Polymer of	% Elements Found					Empirical Formula of Repeating Unit
	C	H	N	O	Others	
Acrylonitrile	59.16	5.61	24.68	10.55	-	$C_3H_3N_1O_{0\cdot4}$
Propionitrile	55.19	7.3	18.56	18.95	-	$C_3H_{4\cdot7}N_1O_{0\cdot8}$
Propylamine	63.23	8.89	18.02	9.86	-	$C_3H_5N_1O_{0\cdot4}$
Allylamine	60.48	7.86	18.43	13.23	-	$C_3H_{4\cdot7}N_1O_{0\cdot4}$
Ethylene	74.13	8.11	-	17.77	-	$C_2H_{2\cdot6}O_{0\cdot4}$
Ethylene/N_2	49.38	6.26	18.51	25.85	-	$C_2H_3N_{0\cdot6}O_{0\cdot8}$
Allene	78.26	8.13	-	13.61	-	$C_3H_{3\cdot7}O_{0\cdot4}$
Allene/H_2O	73.15	8.63	-	18.22	-	$C_3H_{4\cdot2}O_{0\cdot6}$
Allene/N_2	62.68	6.71	17.73	12.88	-	$C_3H_{3\cdot8}N_{0\cdot7}O_{0\cdot5}$
Allene/N_2/H_2O	63.77	7.86	10.79	17.58	-	$C_3H_{4\cdot4}N_{0\cdot45}O_{0\cdot6}$
Acetylene	79.5	5.4	-	15.1	-	$C_2H_{1\cdot6}O_{0\cdot3}$
Acetylene/N_2	64.0	5.8	16.7	13.5	-	$C_2H_{2\cdot2}N_{0\cdot5}O_{0\cdot3}$
Acetylene/H_2O	66.5	7.6	-	25.9	-	$C_2H_{2\cdot7}O_{0\cdot6}$
Acetylene/N_2/H_2O	53.2	6.5	15.7	24.6	-	$C_2H_{2\cdot9}N_{0\cdot5}O_{0\cdot7}$
Ethylene Oxide	72.6	9.0	-	18.4	-	$C_2H_{2\cdot9}O_{0\cdot4}$
Hexamethyldisiloxane	30.0	7.5	-	22.2	40.2 (Si)	$C_{3\cdot5}H_{10\cdot5}Si_2O_2$
Tetrafluoroethylene	27.8	-	-	5.28	66.92 (F)	$C_2F_3O_{0\cdot3}$

[a]Reprinted courtesy of Marcel Dekker, Inc.

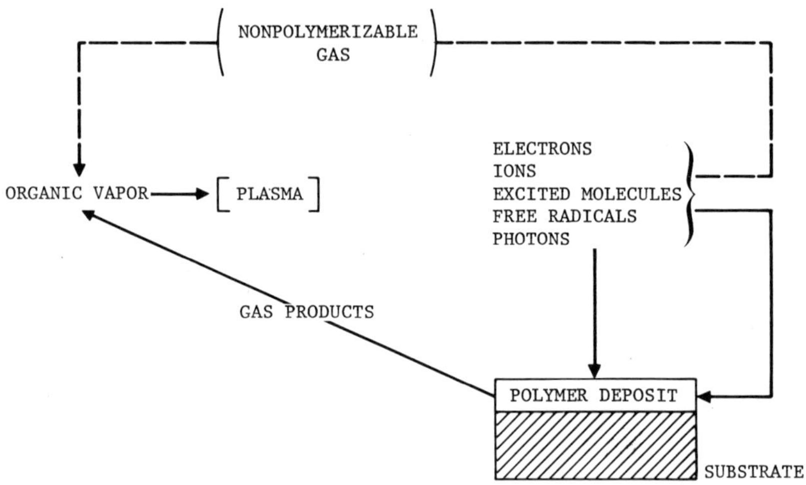

Figure 2-7. Interaction of polymer-forming plasma with a polymer surface (Yasuda, 1976). Reprinted courtesy of Marcel Dekker, Inc.

Table 2-6. Monomers Used to Form Polymeric Films in a Glow-Charge System (Smolinsky, 1968)[a]

Formed Best Films	Formed Films Slowly
Dimethylpolysiloxane	Cyclohexane
Triethylsilane	Hexane
Diethylvinylsilane	Chlorobenzene
Vinyltrimethylsilane	1-Chloropropene
Cyclohexene	1-Chloro-2-methylpropane
Heptene-2	Butanol-1
Formed Very Good Films	Proprionaldehyde
2,5-Dimethyl-2,4-hexadiene	2-Heptanone
Styrene	Not Suggested as Monomers
Valeronitrile	Cyclopentanone
Formed Satisfactory Films	2-Buten-1-ol
Toluene	Ethyl Butyrate
Xylene	Tetrahydrofuran
Benzene	Pyridine
	Tetravinyl Tin
	Perfluorohexane
	Trichloropentafluoropropane
	Did Not Form Films
	Tetrachlorotetrafluoropropane
	Chlorotrifluoroethylene

[a]Reprinted with permission of the American Chemical Society.

Table 2-7. Polymerization Efficiencies of Monomers Deposited on Parallel-Plate Electrodes in a Static System (Bradley and Hammes, 1963)[a]

Monomer	Yield (g/kWh)	Moles (kWh)
Vinyl Ferrocene	300	1.4
1,3,5-Trichlorobenzene	190	1.05
Chlorobenzene	75	0.67
Styrene	69	0.66
Ferrocene	67	0.36
Picoline	65	0.70
Naphthalene	62	0.48
Pentamethylbenzene	61	0.44
Nitrotoluene	55	0.40
Acrylonitrile	55	1.04
Diphenyl Selenide	52	0.20
p-Toluidine	47	0.44
p-Xylene	45	0.42
N,N-Dimethyl-*p*-toluidine	39	0.29
Toluene	38	0.41
Aniline	38	0.41
Diphenyl Mercury	30	0.08
Hexamethylbenzene	28	0.17
Malononitrile	25	0.38
Tetracyanoethylene	19	0.15
Thiophene	13.5	0.16
Benzene Selenol	13.5	0.086
Tetrafluoroethylene	12	0.12
Ethylene	11	0.39
N-Nitrosodiphenylamine	10	0.05
Thianthrene	10	0.046
Acetylene	9	0.35
N-Nitrosopiperidine	9	0.08
Dicyanoketene Ethylacetal	7	0.04
Cyamelurine	6	0.013
1,2,4-Trichlorobenzene	5.5	0.03
Propane	5.2	0.12
Thiourea	4.7	0.06
Thioacetamide	4.4	0.059
N-Nitrosodiethylamine	4.2	0.04
Hexa-n-butyl(di)tin	4.0	0.007
Triphenyl Arsenic	3.4	0.19
Methyl Mercaptan	1	
Carbon Tetrachloride	0	
Hexachlorobenzene	0	
Ammonia	0	

[a]Reprinted by permission of the publisher of The Electrochemical Society, Inc.

Properties of plasma-polymerized thin films are very interesting (Hollahan and Bell, 1979; Seegar, 1977). Films are formed to be amorphous, pinhole-free, highly cross-linked, thermally stable, and have high melting points and low solubility. They undergo surface oxidation when exposed to atmosphere because of trapped free radicals. Usually the film thickness is limited to 2 μ if brittleness is to be avoided. Chemically, they contain considerable unsaturation sites in the form of olefinic bonds and free valences. Carbonyl bonds occur by exposition to the atmosphere. Production of aromaticity was not observed, but this characteristic is preserved when the monomer utilized is aromatic. Structures of polymers obtained in glow are presented in Table 2-8 (Kronick et al., 1969). The electrical conductivity is in the range of 10^{-6} to 10^{-4} mho/cm; the dissipation factor is on the order of 0.025.

Table 2-8. Structure in Glow-Discharge Polymers from Hydrocarbons (Kronick et al., 1969)[a]

Starting Vapor	Functional Groups in Polymers
Pentane	Branches at each pentane molecule, methyl chain ends (–CH=CH–).
Ethylene	(–CH_2–CH_2–CH_2–), methyl chain ends, (–CH=CH), cross-links at saturated carbons.
Butadiene	(–CH_2–CH_2–CH_2–), (–CH=CH–), methyl chain ends, and cross-links at saturated and unsaturated carbons.
Benzene	(–CH_2–CH_2–CH_2–), (–CH=CH–), (–C≡C–) or (–C=C=C–), methyl chain ends and pentyl side groups.
Styrene	Same as benzene, (C_6H_3)–CH_2–.
Naphthalene	(–CH_2–CH_2–CH_2–), (–CH=CH–), (–C≡C–) or (–C=C=C–), methyl chain ends, and 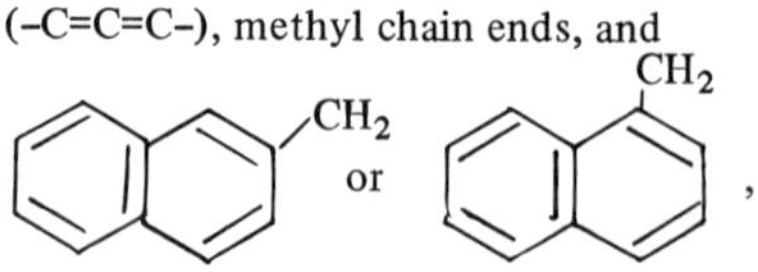 , and 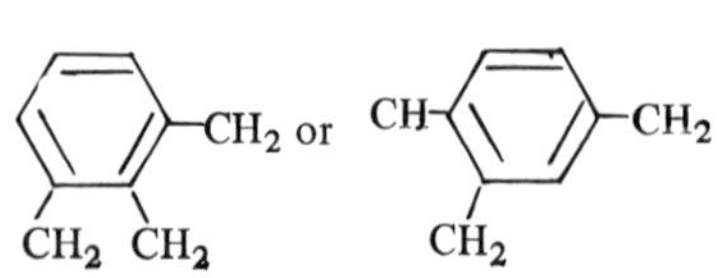

[a]Reprinted by permission of John Wiley & Sons, Inc.

Applications of plasma-deposited films include dielectrics in the electronics industry, protective coatings for metal and glass (Weydeven et al., 1976, 1977; Hollahan et al., 1974), semiconductors and other reactive surfaces and membrane fabrication. We will briefly discuss a few results on membranes produced by plasma polymerization.

By plasma polymerization, one can prepare thin membranes that reject dissolved solids such as ions and organics (semipermeable membranes) not only with conventional, unsaturated organic monomers, but also with saturated monomers. These membranes are thin and yield a high water flux at the same time as a high salt rejection. On the other hand, for conventionally prepared membranes, high permeability is accompanied by low salt rejection. Another feature rendered possible by plasma is the choice of a substrate with the smallest possible compaction during filtration. The membrane themselves are highly cross-linked and do not undergo further compaction. They are deposited in a partial vacuum, i.e., dust-free and dry, and can be stored indefinitely in the dry state, whereas conventional ones must be stored wet. A variety of configurations such as hollow fibers and tubules seem possible in addition to flat sheets. Some reverse osmosis results obtained from various monomers are shown in Table 2-9. Hyperfiltration membranes have also recently been made of allylamine

Table 2-9. Reverse Osmosis Results of Plasma-Polymerized Polymer from Hydrophilic Type A Monomers (Hollahan and Bell, 1974)[a]

Monomer	Substrate	NaCl Concentration (%)	Applied Pressure (psi)	Salt Rejection (%)	Water Flux (gfd)[b]
4-Vinylpyridine	Porous glass	1.2	1200	87-90	0.49-0.72[a]
	Polysulfone	1.2	1200	89-97	1.6-7
	Millipore-VS	1.2	1200	95-99	4-38.0
N-Vinylpyrrolidone	Polysulfone	1.2	1200	91	10.6
4-Picolin	Polysulfone	1.2	1200	98	6.4
4-Ethyl Pyridine	Polysulfone	1.2	1200	98	9.6
4-Picolin	Polysulfone	3.5	1500	96	7.6
4-Methylbenzylamine	Polysulfone	3.5	1500	96	2.2
n-Butylamine	Polysulfone	3.5	1500	94	2.7
4-Vinylpyridine	Polysulfone	3.5	1500	97	4.9
4-Vinylpyridine	Porous glass	3.5	1500	96	0.80[c]
3-5 Lutidine	Polysulfone	3.5	1500	99	12.0

[a]Reprinted by permission of John Wiley & Sons, Inc.
[b]Gallons per foot per day.
[c]Water flux of porous glass tube is approximately 1.0 gfd.

monomer (Bell et al., 1975). The process of forming a semipermeable membrane on a porous substrate is interpreted as the building up of a layer of polymer on the pore walls, which can proceed until closure of the pore is achieved (Yasuda, 1972; Yasuda et al., 1973).

CHAPTER 3

ORGANIC SYNTHESIS THROUGH THE EFFECT OF SINGLET MOLECULAR OXYGEN

INTRODUCTION

The oxidation of organic and biological substrates under the influence of light, oxygen and a sensitizer was the path that led to the discovery of the action of a singlet molecular oxygen. Two types of photosensitized oxygenation reactions are observed:

1. The excited sensitizer initiates a free radical-propagated autooxidation. An example of photochemically initiated autooxidation is the benzophenone-sensitized oxidation of isopropyl alcohol in the presence of oxygen,
2. The reactive intermediate is an electronically excited state of molecular oxygen (singlet oxygen) produced by the transfer of energy from the sensitizer to oxygen to produce a reactive metastable state of the oxygen molecule (Kautsky mechanism). The term "photosensitized oxidation" is synonymous with oxidations performed with singlet oxygen. Kautsky demonstrated that the photoexcitation of a sensitizer in the presence of oxygen led to oxidation of a substrate physically removed from the sensitizer. He postulated that the oxidant was excited molecular oxygen in its metastable singlet state. This explanation was validated more than three decades later when several laboratories showed by spectroscopic methods that triplet aromatic hydrocarbons were quenched by ground state oxygen and that energy transfer to the oxygen gave rise to singlet oxygen (Kaplan 1971).

$$^{1}\text{sensitizer} + h\nu \longrightarrow {}^{3}\text{sensitizer}$$

$$^{3}\text{sensitizer} + {}^{3}\Sigma_g\text{-}O_2 \longrightarrow {}^{1}\Delta_g O_2 + {}^{1}\text{sensitizer}$$

$$^{1}\Delta_g O_2 + \text{acceptor} \longrightarrow \text{acceptor}-O_2$$

Examples of early photooxygenation reactions are:

- photooxygenation of naphthacene to a peroxide (1867);
- photooxygenation of zubrene (1926);
- dye-sensitized oxygenation of ergosterol (1928); and
- synthesis of ascaridole from α-terpinene (1944).

PROPERTIES OF SINGLET MOLECULAR OXYGEN

Oxygen is ground state triplet molecule. Therefore, transitions from the first two excited singlet states, $^1\Delta_g$ and $^1\Sigma_g^+$, are forbidden processes. Thus, these states are expected to have long radiative lifetimes. The $^1\Sigma_g^+$ and $^1\Delta_g$ states are 37 and 22 kcal above the ground state, respectively. The $^1\Sigma_g^+$ state is collisionally deactivated much more rapidly than the $^1\Delta_g$ state. In general, the $^1\Delta_g$ state is believed to be responsible for most of the observed chemistry, although there are cases where it is believed that the $^1\Sigma_g^+$ state is influencing the chemistry. The $^1\Delta_g$ state has both electrons paired in a single orbital, leaving one of the highest orbitals empty. Such a species would be expected to undergo electrophilic-type reactions. Table 3-1 shows the electronic states of molecular oxygen and their properties.

Table 3-1. Electronic States of Molecular Oxygen and Their Properties (Schaap, 1976; Kaplan, 1971; Murray et al.)

State	$^1\Sigma_g^+$	$^1\Delta_g$	$^3\Sigma_g^-$
Energy Above Ground			
kcal/mol	37.5	22.5	0
ev	1.63	0.977	0
Absorption Wavelengths[a], μ			
Micron	633.4 and 703.2	1,070 and 1,261	
Orbital Occupancy Scheme	$\uparrow\ \downarrow$	$\uparrow\downarrow$ —	$\uparrow\ \downarrow$
Light Emitted, cm^{-1}	13,120.9	7,882.4	
Radiative Lifetime at Zero Pressure, sec	7 sec	45 min	
Half-Life at 1 Atmosphere, sec		5×10^{-2}	
Gas-Phase Collisions Survived	$10\text{-}10^2$	10^4	
Lifetime in Solution, sec	10^{-9} to 10^{-12}	10^{-3} to 10^{-6}	
Wall Collisions Survived	10^3	2×10^5 to 3.5×10^4	
Mean Diffusion Path in Thin Films, Å		115 ± 20	

[a]For transition to ground state.

Using a pulse laser technique, Merkel and Kearns (1971) were the first to determine a lifetime of $^1\Delta_g$ $^1O_2$ in solution: 7 μsec in methanol. Table 3-2 gives the lifetimes in various solvents.

Table 3-2. Lifetimes of Singlet Molecular Oxygen ($^1\Delta_g$) in Various Solvents (Schaap, 1976)

Solvent	μsec
H_2O	2
D_2O	20
CH_3OH	7
CH_3CH_2OH	12
C_6H_{12}	17
C_6H_6	24
CH_3COCH_3	26
CH_3CN	30
$CHCl_3$	60
CS_2	200
$CDCl_3$	300
C_6F_6	600
CCl_4	700
CF_3Cl (Freon II)	1000
Methanol	7-9
Benzene 25	26
Benzene d6	36
Ethanol	10
Ethanol 95%	5

METHODS FOR GENERATION OF SINGLET OXYGEN

The methods that have been considered and utilized in investigative work on $^1O_2$ include:

- photosensitization;
- reaction of sodium hypochlorite with hydrogen peroxide;
- thermal decomposition of phosphate ozonides;
- decomposition of epidoxides;
- reaction of potassium superoxide in water;
- self-reaction of sec-peroxy radicals;
- microwave discharges in oxygen-containing gases; and
- light and laser irradiation

These modes of production of $^1O_2$ will now be presented in some detail.

Photosensitization

Historically, this was the first process that led to the awareness of the existence of $^1O_2$. A light-absorbing compound, usually dye, is excited by light in the presence of oxygen and transfers its energy to the oxygen.

Chemical Production

One typical production is by the reaction of sodium hypochlorite and hydrogen peroxide. Whereas triplet oxygen can be directly produced from one electron reaction of hydrogen peroxides, such as

$$(Fe(CN))^{3-} + \text{-O-OH} \longrightarrow (Fe(CN))^{4-} + \text{.O-OH}$$
$$(Fe(CN))^{3-} + \text{-O-OH} \longrightarrow (Fe(CN))^{4-} + \text{.O.O-} + H+$$

singlet oxygen must be formed initially from its heterolytic (i.e., electron pair) decomposition, such as

$$\text{O-O-H} + \text{-O-Cl} \longrightarrow \text{O=O} + \text{H-O} + \text{Cl-}\cdot$$

Only the heterolytic reactions are chemiluminescent. This light emission is due to the rapid conversion of singlet oxygen into triplet oxygen. Alkaline hydrogen peroxide and molecular bromine were also used to generate singlet oxygen according to the reaction

$$\text{O-O-H} + \text{Br-Br} \longrightarrow \text{O=O} + \text{HBr} + \text{Br-}$$

However, the utility of the method using bromine is often lessened by side reactions.

Other chemical sources of singlet molecular oxygen of practical interest include the following:

1. Triphenyl phosphite ozonide, formed at -78° C by the addition of ozone to the phosphite, will decompose at -35° C to yield singlet oxygen (Murray and Kaplan, 1969). Several cyclic phosphite ozonides of much greater stability have been reported by Brennan (1970) and Stephenson et al. (1973).
2. Ozone reactions with organic compounds can produce singlet oxygen. Examples of expected sources are given in Table 3-3 (Murray et al.).
3. Thermal decomposition of 9,10-diphenylanthracene endoperoxide produces singlet oxygen.

It has been suggested that the dismutation reaction of superoxide ion radical yields singlet oxygen (Khan, 1970). Enzymes produce the superoxide anion radical

$$\cdot O_2^- + \cdot O_2^- \xrightarrow{H^+} {}^1O_2 + H_2O_2$$

Table 3-3. Examples of Expected Sources of Singlet Oxygen in Ozone Chemistry (Murray et al.)

$$R_3N + O_3 \longrightarrow R_3N^+ - O^- + O_2$$

$$R_3P + O_3 \longrightarrow R_3P = O + O_2$$

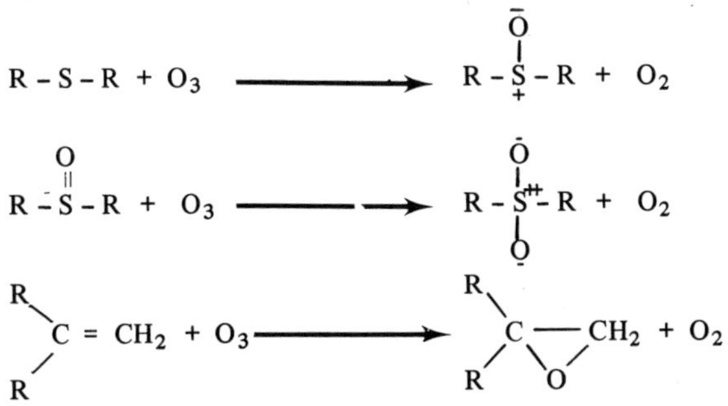

An interesting example where a reaction generates $^1O_2$ is the Russell termination mechanism for sec-peroxyradicals (Horn and Ingold, 1968), although this is not a useful preparative source of $^1O_2$. $^1O_2$ appears to be produced in the base-induced decomposition of peroxyacetylnitrate (Steer et al., 1969):

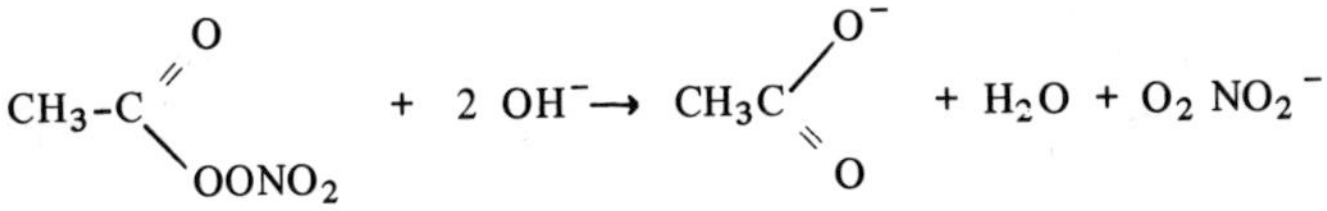

Production by Electric Discharge

It was demonstrated in 1956 by Foner and Hudson that singlet molecular oxygen is produced by passing a stream of oxygen through an electric discharge. It was estimated that 10-20% of the oxygen molecules are in the $^1\Delta_g$ state, based on measurements with a mass spectrometer and a collision-free molecular beam sampling system.

Corey and Taylor (1964) passed oxygen through a similar electric discharge and bubbled it into a solution containing the acceptors to yield endoperoxides. Gollnick and Schade (1973) produced $^1O_2$ by microwave discharges in CO_2, N_2O, NO_2 and SO_2.

Production by Laser Irradiation and by Light

By a fortunate coincidence, the normal wavelength of an He-Ne laser, 632.8 nm, almost exactly coincides with that of the $(^1\Delta_g)_2 \rightarrow (^3\Sigma_g^-)_2$ double-molecule transition (Evans, 1969). Irradiation of 9,10-dimethylanthracene or 1,3-diphenyl isobenzofuran under 2000 psi of oxygen leads to conversion of the acceptors to the usual singlet oxygen products. The output of a Nd-YAG laser at 1065 nm lies within the $(^1\Delta_g) + \text{v} \rightarrow {}^3\Sigma_g^-$ absorption band of the oxygen molecule, corresponding to the single molecule transition (Matheson and Lee, 1970). Irradiation of a solution of oxygen in Freon-113 with this wavelength produces a singlet oxygen that may be trapped by various acceptors.

Photochemical production of $^1O_2$ in benzene-oxygen mixtures is reported by Snelling (1968) and is attributed to quenching of triplet benzene. The light source utilized was low-pressure mercury lamps.

REACTIONS OF SINGLET OXYGEN WITH ORGANIC SUBSTANCES

Reaction types that have been observed in the laboratory include, with alkenes: (1) 1,4-cycloaddition with conjugated dienes to yield end peroxides; (2) an "ene"-type reaction to form allylic hydroperoxides; and (3) 1,2-cycloaddition (with olefins) to give 1,2-dioxetanes (which thermally decompose to carbonyl containing compounds).

Table 3-4 gives the schematics of these three types of reactions with alkenes:

1. oxidation of sulfides to sulfoxides and sulfones;
2. addition to heterocycles (such as pyrroles, furans, oxazoles, imidazoles, thiophenes); and
3. reactions with biological materials such as fatty acids, purines, pyrimidines, DNA, RNA, amino acids (tyrosine, tryptophan, methionine, cystine, histidine) and proteins.

Some examples of these three basic types of reactions with alkenes are now presented in some detail.

1,4-Cycloaddition of Oxygen to Conjugated Dienes

Synthesis of Ascaridole

This was the first dye-sensitized photooxygenation of a simple 1,3 diene, realized by Schenk and Ziegler in 1943. Irradiation of α-terpinene in the

Table 3-4. Typical Singlet Oxygen Reactions with Organic Molecules

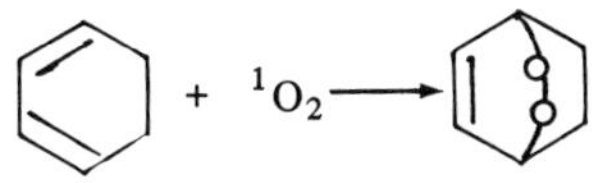

(1)

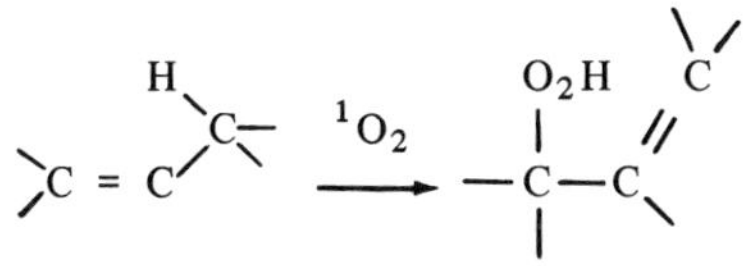

(2)

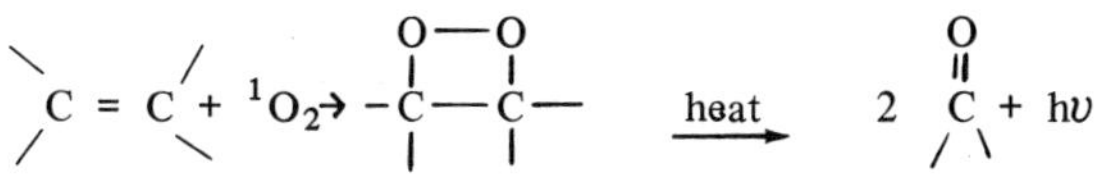

(3)

In the diene reaction an endoperoxide (1) is produced. In the "ene" reaction an allylic hydroperoxide (2) is produced. In the 1,2 cycloaddition, a four-membered ring peroxide (3) is formed.

presence of chlorophyll and oxygen produced a transannular peroxide (nor-ascaridole) from 1,3-cyclohexadiene. The scheme of the reaction is as follows:

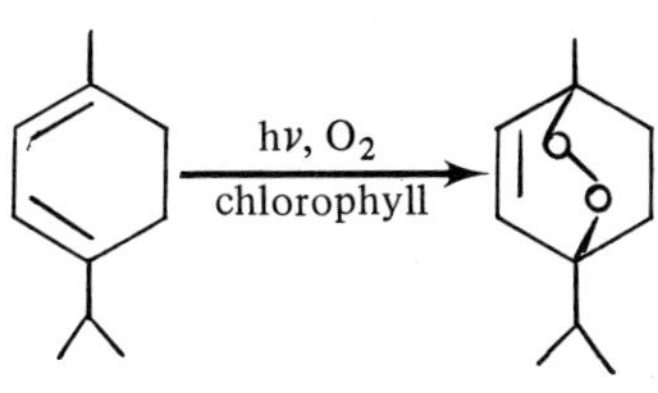

Synthesis of Transannular Peroxide from DPBF (3-diphenybisobenzofuran)

This was achieved by Dufraisse and Ecary in 1946. This peroxide can be considered as the ozonide of 1,2 diphenyl-3,4 benzocyclobutdiene. The scheme of the reaction is as follows:

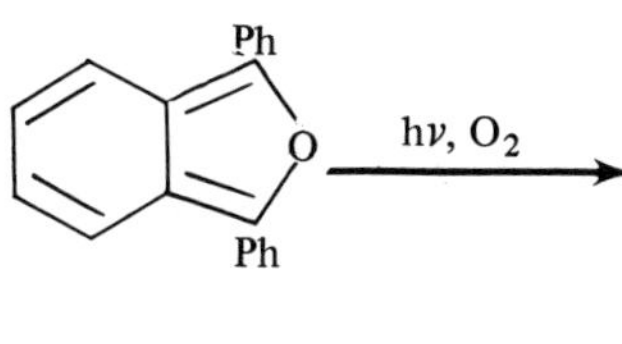

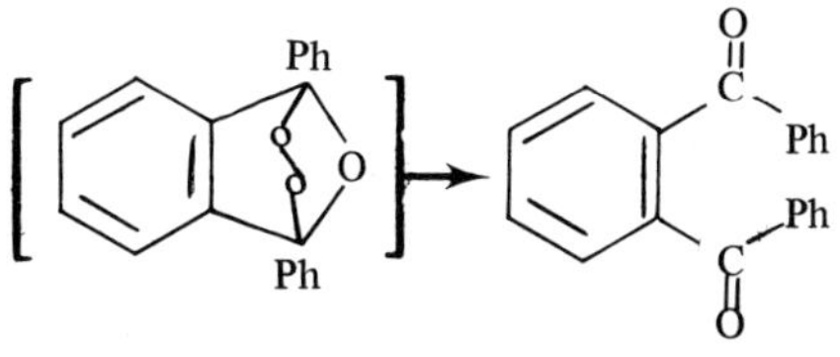

The "ene" Reaction of Singlet Oxygen

This is a type of reaction where the dye-sensitized photooxygenation of alkyl-substituted alkenes gives allylic hydroperoxides, in which the double bond undergoes and allylic shift. The oxygen is added to the same part of the molecule from which the hydrogen is removed.

The reaction is further found to be stereospecific and subject to steric hindrance.

The mechanism of this reaction has been intensively investigated.

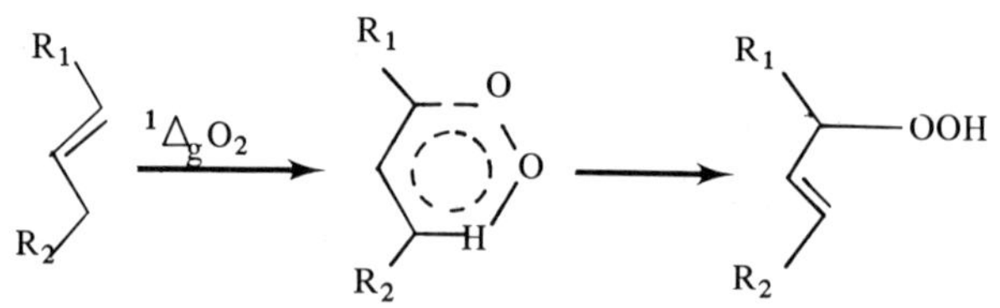

An example of the "ene" reaction is the oxidation of 2-methyl-butene-2:

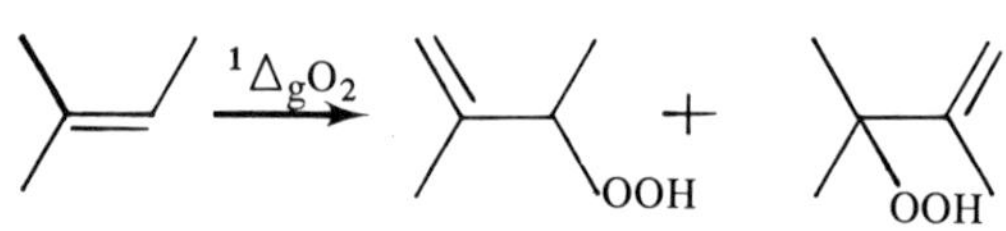

1,2 Cycloaddition to Form 1,2 Dioxetanes

1,2-dioxetanes are four-membered ring peroxides that can be formed from alkenes by singlet oxygen. The first isolation of a stable 1,2 dioxetane was performed in 1969 from α-bromohydroperoxide. Enamines were also found to be cleaved by photooxygenation to form two carbonyl fragments in a partial synthesis of progesterone. Stereospecific *cis* and *trans* addition of a singlet oxygen to vinylene diethers gave isolable 1,2-dioxetanes as products. Other 1,2-dioxetanes were formed by various investigators by addition of singlet oxygen to tetramethoxyethylene, α-β-dimethoxystilbene and adamantylidene-adamantane.

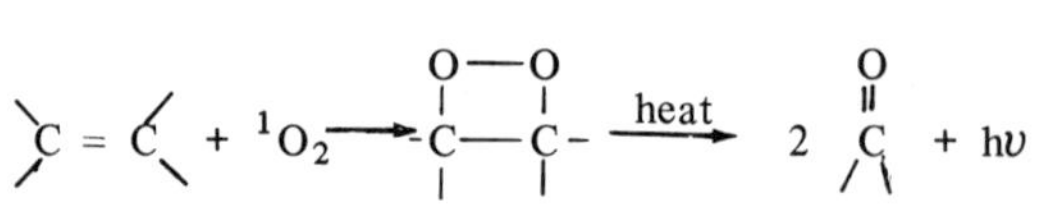

ROLE OF SINGLET OXYGEN IN POLYMER DEGRADATION

Controlled stability of commercial polymers to oxygen is a key problem for their long-term properties. Recent research in this field has shown the role of singlet oxygen in the oxidation of synthetic polymers and the possible importance of singlet oxygen quenching in photostabilization processes, which can now be considered for future individual application (Ranby and Rabek, 1978).

Singlet oxygen in a polymer matrix can be formed in two ways:

- in photosensitizing reactions due to impurities (polyolefins exposed to air can absorb polynuclear aromatics released into the atmosphere from combination processes, which are able to generate $^1O_2$); and
- in energy transfer reactions between excited triplet states of carbonyl-type groups embedded in a chain and molecular oxygen.

In addition, the surface of polymers can be exposed to single molecular oxygen formerly in photochemical reactions occurring in polluted or damp atmospheres surrounding the sample.

Singlet oxygen can be responsible for the initiation mechanism of polymer oxidation and can detach hydrogen atoms to ultimately form water, depending on the polymer type.

Polymers with saturated backbone chain can undergo the following changes:

1. Carbonyl groups are excited by UV irradiation.
2. C-C bonds in excited carbonyl groups are cleaved, resulting in degradation or transfer of energy from excited carbonyl groups to molecular oxygen-forming $^1O_2$ and reaction of $^1O_2$ with vinyl groups to make hydroperoxides.

Polymers with unsaturated backbone chains react readily with $^1O_2$ to form endoperoxides. Singlet oxygen oxidation can be photosensitized by polynuclear aromatic hydrocarbons, quinones and dyes. Some typical oxidations with singlet oxygen are shown in Table 3-5. By contrast, photostabilizers and oxidants can quench singlet oxygen (Wiles, 1978) and retard or suppress polymer degradation.

In principle, the "stabilization" of macromolecules against degradation resulting from $^1O_2$ attack can be based on three concepts:

1. prevention of $^1O_2$ formation in and on solid polymers;
2. deactivation of $^1O_2$ by quenching before it reacts with the polymer; and
3. decomposition or scavenging of the products from the $^1O_2$ with the polymer.

Table 3-5. Some Typical Oxidations With Singlet Oxygen (Kaplan, 1971)

Substrate	Product
Olefins 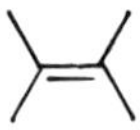	 Allylic hydroperoxide
1,3-Dienes 	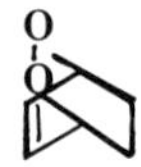 Endoperoxides
Thiophenes 	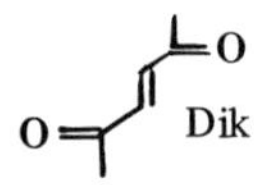 Diketones 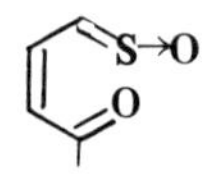 Sulfines
Furans 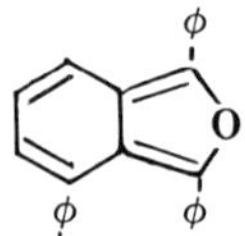	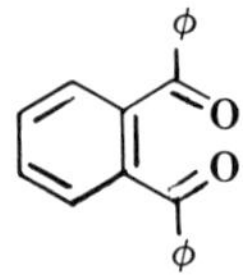 Diketones
Pyrroles 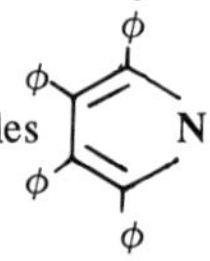	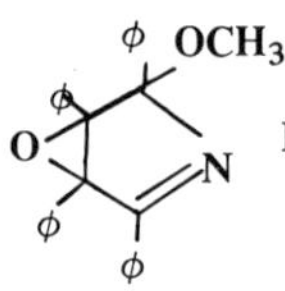 Epoxides (photo-oxidation in CH_3OH)
Enamines 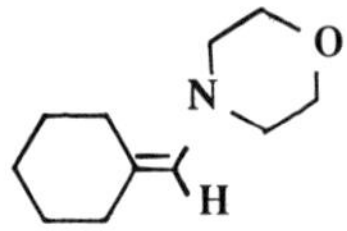	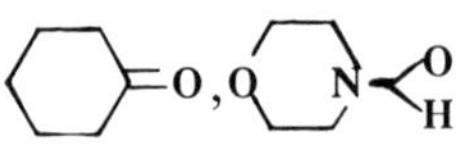 Carbonyl Amide
Oxazoles 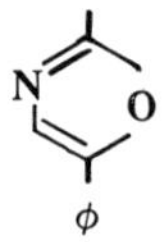	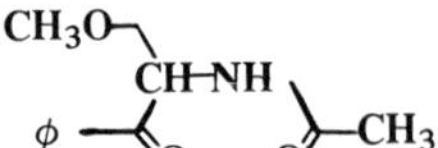 Acetamidoketones
Phenols 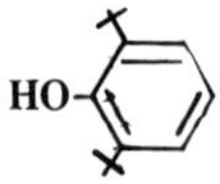	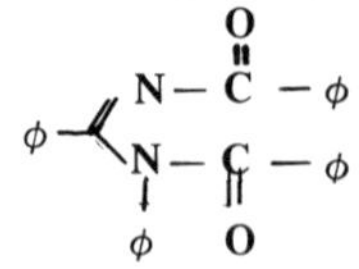 N,N′-Diaroylbenzamidines
Imidazoles 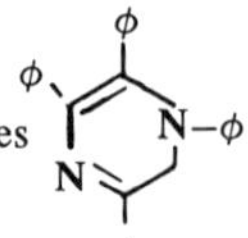	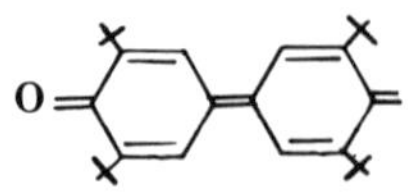 Benzoquinones
Polybutadiene 	 Hydroperoxide

To prevent $^1O_2$ formation, one can reduce the formation of ozone complexes (which decompose giving $^1O_2$), reduce the formation of peroxy radicals, and eliminate UV-sensitive impurities (which is limited by costs).

As $^1O_2$ cannot entirely be prevented from forming, compounds that will deactivate it back to $^3O_2$ in the liquid phase can be used. Such deactivators (quenchers) are amines, sulfides, phenols, nickel chelates and β-carotene (the most efficient of all, but photo- and thermally unstable). Table 3-6 gives examples of UV stabilizers used for prolonging the life of polypropylene films that are also good quenchers of $^1O_2$.

Table 3-6. Comparison of Polypropylene Film UV Stabilization (hours) and ^1O ($^1{}_g$) Quenching Rate Constants (k_q) (Wiles, 1978)

Additive	PP Lifetime[a] (hr)	k (M^{-1} S^{-1q} x 10^{-8})[b]
None	110	-
β-Carotene	---	85[c]
Ni(II)diisopropyldithiophosphate	430	54[c]
Fe(III)diisopropyldithiocarbamate	---	38[c]
Ni(II)diisoprophldithiocarbamate	1310	34[c]
Ni(II)*bis*[2-hydroxy-5-methoxyphenyl-N-(n-butyl)aldimine]	380	24[d]
CO(II)diisopropyldithiocarbamate	---	19[c]
Ni(II)*bis*[2,2′-thiobis-(4-tert-octyl) phenolate]	250	1.3[d]
Ni(II)acetylacetonate	250	0.75[c]
Zn(II)diisopropyldithiocarbamate	186	<0.1[c]
2-Hydroxy-4-dodecyloxybenzophenone	350	<0.01[d]

[a]Time to brittle failure of 30-μm films containing 0.1 wt % of additive: Atlas carbon-arc Fade-Ometer irradiation.
[b]Additive (quencher) concentration 1 x 10^{-5} to 10^{-3} mol^{-1}.
[c]In methylene chloride solution.
[d]In isooctane solution.

As it is probable that some $^1O_2$ will still react with the polymer, one will have to prevent the reaction products (mainly the hydroperoxides and the alkoxy, alkyl and peroxy radicals) from doing damage by decomposing them. Scavenging of such radical intermediates can be achieved by using some photostabilizers (nickel chelates) and a hydroxybenzophenone, for instance. Efficient photostabilization of polyolefins requires this kind of free radical trapping.

IMPROVING SELF-ADHESION OF SYNTHETIC TIRE RUBBER BY MEANS OF A SINGLET OXYGEN REACTION

The following appears to be one of the very first applications of singlet oxygen knowledge to industrial purposes. Self-adhesion is where two specimens of the same material adhere to each other when brought into contact. In tire manufacture, rubbers are required to possess this property before vulcanization.

EPDM is a synthetic rubber composed of ethylene, propylene and a third monomer. It has a much higher UV and ozone resistance than natural rubber, but an insufficient level of self-adhesion (tack).

It has been found that with suitable choice of third monomer and additives and exposure to visible light, a very high self-adhesion is obtained. In this development, the choice of the third monomer was guided by the requirement that it react well with $^1O_{2+}$ and the choice of the extender oil by the requirement that it be a good photosensitizer. As a check that the self-adhesion property was really caused by a singlet molecular oxygen reaction, a known $^1O_2$ quencher was added, resulting in a considerable reduction of self-adhesion (Wolters et al., 1978). It is believed that the formation of hydroperoxides by singlet oxygen leads to a highly cross-linked surface, which, together with the increased surface polarity, should be responsible for the desired effect.

APPLICATION OF KNOWLEDGE ON SINGLET OXYGEN REACTIONS IN THE PAPER AND PULP INDUSTRY

In recent years, lignin-retaining bleaching has been developed in this industry to the point that bleached mechanical pulp is now applied as a substitute for bleached chemical pulp. The advantages are better utilization of the wood raw material and a large reduction of air and water pollution; however, such pulps have a tendency to yellow, and ways have to be found to stabilize them against the influence of exposure to daylight. The light-induced yellowing of lignin-rich pulp proceeds in three steps (Gellerstedt et al., 1978):

1. absorption of light by a number of structural groups in lignin (double bond end groups of cinnamyl alcohols and cinnamaldehydes, carbonyl groups conjugated with aromatic rings), whereby singlet oxygen is produced when these groups in their excited states transfer energy to oxygen;
2. formation of phenoxy radicals by hydrogen abstraction (detachment); and
3. reactions between the phenoxy radicals thus created an oxygen with the formation of quinones in color-forming reactions.

In addition, lignin-retaining bleaching of mechanical pulp is usually done with hydrogen peroxide in alkaline solution. It has been observed that singlet oxygen was already formed by disproportionation of the hydrogen peroxide in the alkaline solution. Singlet oxygen can attack the ionized phenolic groups in lignin, found to be in alkaline solution, a more powerful oxidant of phenols than ordinary oxygen.

On the basis of these observations, one can conclude that singlet oxygen is an important oxidant in reactions of interest to the pulp and paper industry and, therefore, its properties can be utilized to the advantage of the industry.

POTENTIAL APPLICATIONS OF SINGLET OXYGEN

Currently, singlet molecular oxygen does not appear to be used in any industrial process in the U.S. It is only used in the research laboratory. Nevertheless, it is possible to point out some areas where $^1O_2$ could be utilized, such as the following for organic synthesis and biomedical applications.

- fine chemical synthesis of flavors, perfumes by photochemistry in the cosmetic industry;
- propylene oxide synthesis from propylene (a multibillion dollar a year industry; however the chemical industry does not seem to consider electric methods for this application as yet because they give no economy of scale);
- preparation of escaridole;
- reaction with polymers (Kaplan et al., 1972);
- superoxide anion radical mechanism for damage in cells;
- treatment of melanin loss in skin by singlet oxygen;
- dye-sensitized photooxygenation of kryptopyrole (Lightner et al., 1973a); and
- *in vitro* chemistry of biological molecules (Truscott et al., 1973).

RESEARCH AREAS CURRENTLY ACTIVE IN SINGLET MOLECULAR OXYGEN SCIENCE

The following areas appear to be those where most of the investigative work is now proceeding (Wasserman, 1978).

- singlet oxygen electronic structure and photosensitization;
- gaseous singlet oxygen;
- chemical sources of singlet oxygen;
- solvent and solvent isotope effects on the lifetime of singlet oxygen;
- quenching of singlet oxygen;
- 1,2-Cycloaddition reactions of single oxygen;
- 1,2-Dioxetanes;

- ene-reactions with singlet oxygen;
- reactions of singlet oxygen with heterocyclic systems;
- oxidations of electron-rich aromatic compounds;
- role of singlet oxygen in the degradation of polymers; and
- biological roles of singlet oxygen.

CHAPTER 4

INVESTIGATIONS ON APPLICATIONS OF LOW-TEMPERATURE PLASMAS TO ORGANIC SYNTHESIS

INTRODUCTION

The interest in investigating glow discharges for organic synthesis is twofold: (1) to create new compounds; and (2) to synthesize in a single reaction substances usually needing several steps.

To date, except for the synthesis of ozone, low-temperature plasmas have not been applied in industry for chemical synthesis. This chapter presents the basic results obtained in laboratories, based on the classification derived by Suhr, who is one of the leaders in research in this area (Hollahan and Bell, 1974). Only those reactions that are highly selective and give high yields can be of practical interest and will be mentioned.

Four major types of reactions in glow discharges are of value to preparative chemistry:

1. generation of atoms and radicals;
2. isomerization (including migration of substituents and rearrangements of cyclic compounds);
3. elimination of atoms and small groups; and
4. bimolecular reactions.

GENERATION OF ATOMIC GASES AND THEIR REACTION WITH ORGANIC COMPOUNDS

The major gases of interest have been hydrogen, oxygen and nitrogen. Either the atoms are generated in the discharge and the organic material is injected behind the plasma zone or the organic substance is fed into the discharge together with the molecular gas, which will be atomized by the discharge. Table 4-1 indicates the reaction paths of various organic compounds and the end products for several feed gases.

ISOMERIZATIONS, INCLUDING MIGRATION OF SUBSTITUENTS AND REARRANGEMENT OF CYCLIC COMPOUNDS

Plasma isomerizations are limited to molecules without highly reactive groups. In addition to *cis-trans*-isomerization (preparation of the *cis*-isomer from the *trans*-isomer), migration of substituents (radicals) is observed. Another kind of isomerization found is rearrangement of cyclic compounds, whereby a ring is transformed into a chain (a process shared with side-reactions whereby a fraction of the starting material goes directly into polymers). Table 4-2 shows typical examples of these reactions.

ELIMINATIONS, INCLUDING STABILIZATION OF PRODUCTS BY CYCLIZATION

Elimination reactions occur where collisions with low-energy electrons in the plasma eliminate small groups or atoms from the molecular without altering the rest. Major reactions of this type are dehydrogenation, decarbonylation, decarboxylations, formation of unsaturated compounds and synthesis of 5- or 6-membered rings. The remaining species often undergo stabilization reactions (cyclization, ring contraction, dimerization). Table 4-3 indicates the most promising reactions and their products with typical yields.

BIMOLECULAR REACTIONS

All reactions investigated in the preceding sections have been monomolecular; we will now review reactions involving two or more molecules reacting with each other in the plasma. In some cases, the reacting molecules can be the products or fragments of one starting material. In other cases, two types of molecules can have been introduced at the start. These reactions require more involved equipment and are more dependent on careful optimization of all discharge conditions. The reactions observed include dimerization, trimerization, oligomerization of original starting material and reactions between different original compounds or products thereof with one of the original compounds. Table 4-4 lists examples of such reactions.

Table 4-1. Glow Discharge Generation of Atoms and Reactions with Organic Compounds

Feed Material	Reacting With	Effect	To Produce (Result)
Hydrogen	Saturated compounds	Abstracts (detaches) other atoms	Organic radical
	Unsaturated compounds	Addition of hydrogen	Organic radical
Oxygen	Saturated compounds		
	Olefins, epoxides	Abstracts atoms	Radicals
	Propylene (passed through discharge with oxygen)	Forms three-membrane rings	Carbonyl compounds
Nitrogen	Most organic compounds	Destruction	CHN, NH_3, low-molecular-weight gases
	Some dienes	1,4 Addition of nitrogen atom to diene	Some pyrrole
Simple molecules (H_2, CH_4, NH_3, H_2O, CO, CO_2)	Reacting together	Synthesis	Aminoacids and oxygen- and nitrogen-containing compounds (low yield, nonspecific reactions)

Table 4-2. Isomerization-Type Plasma Reactions

Nature of Reaction	Nature of Starting Material	Nature of Product	Observations
cis-trans Isomerization	Olefins		
	trans-Stilbene	*cis*-Stilbene	Yield 90% at low power levels, less at high levels in favor of phenanthrene, 30 g/kwh.
	trans-Stilbazol *trans*-Crotonitrile *trans*-Cinnamonitrile	*cis*-Stilbarzol, *cis*-crotonitrile, *cis*-Cinnamonitrile	Fragmentation to acetylene dominates.
Migration of Substituents	Aromatic ethers		
	Alkylaryl ethers	Alkylphenols	80% at low levels, ortho- and para-isomers formed, partically no metaisomer, some fragmentation to phenol.
	Alkyphenyl ethers		
	Isopropylphenyl	Isopropylphenols	Same pattern, more phenol, some cresols yield above 50%, rest phenol.
	Naphthylmethyl ethers	Naphthols	
	Vinyl ethers	Butanal and acetaldehyde	
	Nitrogen compounds (amines)		
	N-alkylanilines	Toluidines	Amilines from side-reactions.
Rearrangement of Cyclic Compounds	Cyclo olefins	Aromatic compounds	Isomerize readily
	Cycloheptatriene	Toluene	
	Cyclo octatoluene	Styrene	
	Methylindole	Quinoline	
	Nitrogen-containing heterocyclics		
	Pyrrole	Crotonitrile	Isomerize readily ring to chain.
	Indole	Benzylcyanide	Isomerization, whereby a considerable fraction of starting material is also converted to polymers.
	Aniline	Unsaturated nitriles	

Table 4-3. Elimination Reactions, Including Stabilization Products of Intermediates

Nature of Reaction	Nature of Starting Material	Nature of Product	Observations
Dehydrogenation	Ethylbenzene	Styrene	Highly selective only on molecules with only one reactive site or one bond that is considerably than all the others.
	Acenaphtene	Acenaphthylene	
	Tetraline	Naphthalene	
	Indane	Indene	
	Stilbene	Phenylacetylene	
	β-Phenylethanol	Styrene	
	β-Bromoelhylbenzene	Styrene	
Decarbonylation (elimination of CO)	Hydrocarbons		
	Benzaldehyde	Benzene	Good yield, if low energies, intermediate benzoil radicals are formed.
	Ketones		
	Acetophenone	Toluene	
	Benzophenone	Diphenyl	
	Benzil	Biphenyl	Almost 100%
	Isotyanales		Form intermediates which fragment before stabilizing.
Decarboxylation (elimination of CO_2)	Carboxylic acids and their esters		
	Benzoic acid	Benzene	Almost 100%
Decarbonylation with Stabilization of Intermediates by Ring Contraction	Bicyclic or trycyclic ketones		
	Fluorenone	Biphenylene	Yield 99%
	Camphor, norcamphor	Trimethylbicyclohexane	Yield 60%

Table 4-3, continued

Nature of Reaction	Nature of Starting Material	Nature of Product	Observations
Decarbonylation with Stabilization of Intermediates by Ring Contraction, continued	Aromatic hydroxy compounds		
	β-naphthol	Indene	Yield of α-isomer 70-80%
	g-anthrol	Fluorene	
Formation of Unsaturated Compounds from Cyclic Anhydrides	Malic anhydride	Acetylene	
	Bicyclic anhydrides	Cycloolefins	
	Cyclodienes	Aromatic systems	Elimination of oxygen.
Elimination Reactions Yielding Intermediate Stabilizing by Cyclization	Aromatic substances		
	Two aromatic rings linked by carbon or hetero atom	Five-membered rings (duranes)	Simple method for synthetizing and heterocyclics, yields 25%
	Two substituted biphenyls	Five-membered rings	Yields 60-80%
	Phenolacetates	Five-membered rings	Yields 20%
	Two phenyl rings separated by two or three bonds	Six-membered carbo and heterocyclic rings	
	Nitrodephenylamine	Phenoxacine	Yields 45%

Table 4-4. Bimolecular Reactions

Nature of Reaction	Nature of Starting Material	Nature of Product	Observations
Dimerization, Trimerization	Aromatic compounds		
	Benzene	Diphenyl	Fairly high yield
	Benzene	Diphenyl and terphenyl	(at higher converstion rates).
	Methyl compounds	Methyl dimerization products	Exceptionally high yields.
	Toluene	Dibenzyl	
	Polymethyl benzenes	Diarylethanes	
	Alkyl aromatics	Diarylethanes	
	Benzyl alcohol, benzyl halides	Dibenzyl	
	Unsaturated compounds		
	Propene	Mixture of hexadrene and hexenes	Dimerization of original molecules and dimerization of fragments occur in parallel.
Oligomerization (involving a larger number of molecules associating)	Acetylene	Products with 6,8 or 10 carbons	If surface of reactor walls greatly enlarged.
Plasma Reactions Between Two Different Compounds	Propene and oxygen	Propylenoxide	Only very high low oxygen content to avoid explosion.
	Styrene and oxygen		
	Benzene and oxygen	Phenol	
	Benzene and ammonia	Aniline	
Reactions Where Product Reacts With Another	Phthalic anhydride and hydrogen		The various intermediates are trapped differently by the different additives (H, C_2H_2, $C^9{}_2$), and some end products are enhanced.
	Phthalic anhydride and acetylene	Phenilacetylene	
	Phthalic anhydride and carbon disulfide	Dibenzylthiophene	

PROSPECTS FOR SYNTHESIS APPLICATIONS

Why there is as yet almost no use of low-temperature plasmas in commercial industrial operations for chemical organic synthesis of bulk products is because:

1. they produce very small quantities of material;
2. there are high starting costs; and
3. there are high electrical energy costs.

Because of this, low-temperature plasma preparation of organic chemicals has only been used in the laboratory for samples in the subgram range. It has been tried for hydrazine, but not pursued because of low yields. The product would have to be either extremely valuable or not produceable by other means to make plasma preparation economically effective.

There is one example in inorganic chemistry where low-temperature plasmas are being examined for bulk production. It is in the manufacture of silicon dioxide fiber to produce a preform from which the optics used as light guides will be drawn for kilometers. The Philips Laboratories in Aachen, Germany are investigating a process whereby silicon dioxide is plasma-deposited as a film having a high refraction index inside a short quartz tube. Here it is sintered on the surface before the tube is shrunk to give the preform. It is not yet established that the process is better than the conventional chemical vapor deposition. The conventional process deposits all silicon tetrachloride doped with germanium, which is then oxidized inside the tube.

The great interest in low-temperature plasmas is in the semiconductor industry. Much of the low-temperature plasma treatment equipment is custom made and goes beyond that which equipment manufacturers deliver. The low-temperature plasmas have revolutionized the semiconductor industry because of their ability to do fine pattern delineation in a wafer. They can do submicron features. Although presently only micron and larger sizes are used, it is expected that submicron sizes should be in commercial production within a few year. Apart from their fine pattern delineation ability, low-temperature plasmas replace liquid etching steps in wafer production. In the near future all steps in the production of wafers will be by dry processing, i.e., utilizing low-temperature plasmas.

All units used in laboratory as well as in industrial applications are very small. Since no large-scale units for low temperature discharges are on the market, these techniques could not be introduced into industries that have to handle large surfaces or quantities. This could change in the future. On the basis of experience with high-powered lasers, new discharge techniques are being developed that allow the scaling-up in reactor dimensions, operational pressure and mass throughput. These systems have not yet been tested for application in chemistry. Hopefully we shall soon know whether these setups are as good as they look.

If the tests come out positive, there will be a number of about 10-100 processes which could be introduced into industry or for which the study of industrial possibilities appears rewarding (Suhr, 1978).

VOLUME AND GROWTH OF THE LOW-TEMPERATURE PLASMA DEVICE INDUSTRY

Before 1977 there was relatively little demand for low-temperature plasma devices (Lindsay, 1978) and total annual sales of equipment manufactured by U.S. companies totalled $2-3 million. In 1977, sales reached $10-12 million and in 1978 sales were expected to double to $20-25 million (Table 4-5). The market is expected to grow exponentially at this rate for some time beyond 1978. All sales figures refer to equipment manufactured in the U.S. Some is exported to Japan, but very little to Europe where the market is not yet growing. Nearly the entire market is for semiconductor manufacture applications. The rest (surface treatments) is still very small but could grow rapidly.

Unit prices for low-temperature plasma discharge equipment range from $2,000-66,000, an average price being $18,000-20,000 across the industry. This gives 500-600 units sold in 1977 and 1,000-1,250 expected to be sold in 1978. An average for the power rating of a unit is 250 watts, giving a rough estimate of 125-150 kW installed in 1977 and 250-317 kW installed in 1978.

Table 4-5. Industrial Markets for Low-Temperature Plasma Devices Manufactured by U.S. Companies (Lindsay, 1978)

	Year			
	1972	1976	1977	1978
Annual Sales, $ million	2	3	10-12[a]	20-25[a]
Unit Price Range, $1000			2-66	2-66
Unit Average Price, $1000	20	20	20	20
Approximate Units Sold Annually	100	150	500-600	1000-1250
Average Unit Power, kW	0.250	0.25	0.25	0.25
Approximate Annually Installed Capacity, kW	25	37.5	125-150	250-317

[a]Abrupt rise due to semiconductor applications.

CHAPTER 5

SURVEY OF PATENTS IN LOW-TEMPERATURE PLASMAS

Survey of Patents (Corona Effect)
Equipment/Procedures

Patent Number	Date of Issue	Inventor	Assignee	Title	Abstract
3,996,122	12/7/76	Frank Eugene Lowther, Severna Park, MD	Union Carbide Corp. New York, NY	Corona Reaction System	A system for subjecting gas to high-voltage corona wherein a gas is sequentially (serially) exposed to a plurality of corona generation zones. In a preferred system, the gas is conducted through several corona generators connected in series by gas conduits, and the gas within the system is cooled subsequent to each exposure to corona.
4,013,415	3/22/77	Igor Sergeevich Burov, ultisa, Yakuba Kolosa, 39, kv. 28 Vladimir Vladimirovich Bysjuk, ultisa Ostrovskogo, 27, kv. 56 Alfred Lvovich, Mosse, Leninsky prospekt, 72a, kv. 66 Leonid Stepanovich Shkurko, ultisa Pulikhova, 35, kv. 1 Vyacheslav Andreevich Vashkevich, ultisa Mendeleeva, 7, kv. 71 all of Minsk, U.S.S.R.		Plasma-Chemical Reactor for Treatment of Disperse Materials	A plasma-chemical reactor for treatment of disperse materials comprises a vertical working chamber. The top portion of the working chamber includes means to feed the disperse material into the reactor. The working chamber is provided with means for feeding and plasma-heating process gas, said means being a number of plasmatrons the nozzles of which are arranged along the periphery of the working chamber bottom portion and directed essentially at a right angle to the working chamber vertical axis. The reactor also includes means for evacuating gaseous products from the working chamber and a bin located under the working chamber and communicating with the bottom portion thereof.

4,038,165	7/26/77	Frank Eugene Lowther, Severna Park, MD	Union Carbide Corp. New York, NY	Corona Reaction Method	Corona discharge reactions are efficiently conducted by passing a reactant gas through a corona discharge gap in which pulsed electrical energy is dissipated to accelerate electrons to a productive energy level, and eliminating gas ions from the gap to minimize nonproductive energizing of the ions prior to additional corona discharge. In one preferred practice of the process, a selective ion neutralization potential is applied to eliminate the nonproductive gas ions between corona pulses. In other preferred practices, a reactant gas exposed to corona is removed from the corona zone after limited exposure to corona to permit recombination of the heavy gas ions prior to additional exposure to corona and/or an ion neutralizing component is added to the reactant gas stream.
3,755,621	11/27/73	Georges J. Gorin Berkeley, CA	LFE Corporation, Waltham, MA	Gas Reaction Apparatus	Apparatus for inducing an efficient reaction between a gaseous plasma and a nongaseous material at relatively low ambient temperatures. The reaction vessel includes means for supporting a sample dish or boat containing the material and for enabling the agitation of the boat to enhance the rate and completeness of the decomposition of the material into its inorganic constituents.

3,619,403	11/9/71	Georges J. Gorin Berkeley, CA	LFE Corp., Waltham, MA	Gas Reaction Apparatus	A gas, such as oxygen, is fed to a cylindrical reaction chamber to which an intense high-frequency electromagnetic radiation field is applied. As the gas enters the reaction chamber it is ionized by the high-frequency electromagnetic field. The ionized gas reacts with a non-gaseous material inserted within the chamber to decompose the material. The resultant products, together with any unreacted species of the gas, are withdrawn from the chamber by a vacuum pump. The reaction chamber is constructed to provide a uniform distribution of the gas throughout the reaction zone of the chamber.
3,758,874	9/11/73	Ethan D. Hoag Boston, MA	Avco Corp., Cincinnati, OH	Corona Discharge Laser	A method of, and apparatus for, producing laser action in and transverse to a flowing gas over a wide range of flow and pressure conditions by an electrical discharge orthogonal to both the flow and lasing direction, which is stabilized by electrode segmentation and ballasting.
3,258,402	6/28/66	P. T. Farnsworth Fort Wayne, IN	ITT Corp.	Electric Discharge-Device for Producing Interactions Between Nuclei	The present invention relates to a space charge device, and more particularly, to a space charge device having utility for producing nuclear reactions. In producing nuclear reactions, this invention utilizes unique apparatus for creating an electric field in space within which charged nuclear particles are oscillated at a sufficient velocity that resulting collisions of particles produce nuclear reactions.

3,526,268	9/1/70	Myron Robinson Highland Park, NJ	Research-Cottrell, Inc., New Jersey	Corona Discharge Heat Transfer	A corona discharge established on a heated wire immersed in a cooler gas disrupts the boundary layer about the wire, thereby enhancing convective heat transfer from the wire to the gas. Under circumstances of other wise natural con-ventions from a 500° F. Wire, for example, corona is observed to raise the heat-transfer rate fourfold. In forced flow at velocities up to at least 90 heat-transfer enhancement due to corona is less marked, but still significant.
	8/12/75	Robert I. Tenney Deerfield, IL	Ozone, Inc. Deerfield, IL	Control System for Corona Discharge Ozone-generating Unit	A control system for an ozone-genera-tions system, that is responsive to the temperature of the ozonized gas by interrupting the coupling of high voltage to the electrodes of an ozone-generating unit. This happens when the temperature at which said ozonized gas rapidly disso-ciates, the circulation of the gas to be ozonized through said unit continuing during the interruption of high voltage. Simultaneously, a timer sequencing por-tion automatically initiaties circulation of air through the ozone generating unit and automatically terminates the coupl-ing of high voltage to said electrodes of the ozone generating unit a given period prior to termination of air circulation where ozonization is to be terminated.

3,896,347	7/22/75	Peter Charles Gelfand Lebanon, PA	Envirotech Corp. Menlo Park, CA	Corona Wind-Generating Device	A device that generates a corona wind that circulates about the support rods for the emitter wires in a electrostatic precipitator to discourage accumulation of oil and dust particles on the insulators from which the support rods are suspended. It includes a ground shell which spacedly surrounds the support rods within the shell. These are spatially arranged to cause the circulating gaseous flow within the shell, which carries entrained particulates to collection on the interior walls of the shell.
3,496,092	2/17/70	James C. Fraser Schenectady, NY	General Electric Co., NY	Solid-State Corona Generator for Chemical-Electical Discharge Processes	A solid-state corona generator energized from a commercially available low-voltage source comprises an autotransformer and full wave diode bridge reactifier for supplying controllable DC potential to a high-frequency series capacitor commutated inverter in the center-tapped supply circuit configuration employing SCR thyristors. In the load circuit is a high-voltage transformer applying 10,000-15,000 volts to a corona reactor connected across the secondary winding which provides a substantial portion or all of the commutating capacitance and is the load during corona discharge. The commutating capacitance is tunable by means of a bank of tuning capacitors connected across the transformer primary winding.

3,836,786	9/17/74	Frank E. Lowther Severna Park, MD	Purification Sciences, Ind., Geneva NY	Dielectric Liquid-Immersed Corona Generator	A plurality of air-tight corona generating cells of a corona generator are supported in a liquid-tight housing where they are immersed in a dielectric liquid such as transformer oil for preventing arcing between the edges of the two electrodes of each corona generating cell. The dielectric liquid is in heat exchange relationship with the external surface of each electrode and is circulated betwen a cooler and the housing for cooling the electrodes. A center support disc located between the electrodes of a cell unexpectedly increases the ozone yield.
3,831,060	8/20/74	Walter Spengler CH, 4105 Biel-Benken Switzerland		Corona Discharge Device	Corona discharge equipment for poorly conducting material is characterized in that one of a pair of electrodes is a rigidly mounted conductor having discrete discharge points spaced along it while the other electrode of the pair is a conductor, the resilient support of which ensures the maintenance of a substantially constant distance between the electrodes of the pair and between the other electrode and the surface to be treated.
3,899,683	8/12/75	Frank Lowther Severna Park, MD	Purification Sciences, Generva, NY	Dielectric Liquid-Immersed Corona Generator	A plurality of air-tight corona generating cells of a corona generator are supported in a liquid-tight housing, where they are immersed in a dielectric liquid such as transofrmer oil for preventing arcing between the edges of the two

					electrodes of each corona generating cell. The dielectric liquid is in heat exchange relationship with the external surface of each electrode and is circulated between a cooler and the housing for cooling the electrodes. A center support disc located between the electrodes of a cell unexpectedly increases the ozone yield.
3,942,093	3/2/76	Frank E. Lowther Severna Park, MD	W. R. Grace & Co. New York, NY	Multiple Corona Generator System	A solid-state power supply system for multiple corona generators, in which a series of bridge-connected corona generators share silicon-controlled rectifiers (SCR). This system provides means by which symmetrical high-frequency power may be supplied to a plurality of corona generators while using a minimum number of SCR devices.
2,892,956	6/30/59	Vincent Vodicka S. Euclid, OH	General Electric Co., NY	Electric Discharge Lamp thereof	This invention relates to electric discharge lamps and the manufacture thereof, particularly to fluorescent lamps of the type in which a fluorescent coating is excited into luminescence.
3,677,931	7/18/72	L. R. O'Hare Albuquerque, NM		Corona Cell for Nitrogen and Other Reactions	A concentric-electrode type of electric chemical reactor cell using high voltages of ac current between a central wire and an outer insulated metallic cylinder, such as in an ozone generator, is adapted for achieving higher energy gas state chemical reactions by the use of higher voltages and fields, thereby producing chemically reactive ionized species which are

					unattainable and undesirable from the standard ozone generator. The ac voltage between the electrodes is provided with a dc bias of such amplitude as to ensure periods of time when there is a greater number of ions in the cell having a charge of the same electric sign as that sign of the central ionizing wire. This has the chemical advantage of reacting that ionized species when the reacting species is more chemically reactive in its ionized form than in its electrically neutral form. Finally, the capacative reactance of the cell is made part of a resonant circuit with the secondary of the current supply transformer to ensure a favorable power factor and efficient operation.
3,900,735	8/19/75	Helmut Jahn Frankfurt am Main West Germany	Hoechst Aktiengesell-schaft, Frankfurt am Main West Germany	Corona Discharge Apparatus	Corona discharge apparatus having a plurality of discharge elements provided with a plurality of resistors for connecting respective discharge elements (or groups of discharge elements) to a voltage source to result in more uniform corona discharge from the discharge elements.
3,899,682	8/12/75	Frank E. Lowther Severna Park, MD	Purification Sciences, Inc., Geneva, NY	Corona Reactor Method and Apparatus	An electrode for use as one of a pair of spaced apart electrodes in a corona reactor cell having a corona discharge chamber between the interior surfaces of the electrodes. One electrode is a flat,

rectangular, planar plate with a continuous peripheral edge that curves away from the plane of the plate in a direction away from the interior surface thereof. The edge includes the entire peripheral edge of said one electrode and has a uniform width and height. The electrode has a ceramic dielectric coating covering the entire area of its interior surface. The electrode can include one or a pair of passageways extending transversely completely there-through, in which case the electrode includes a peripheral wall of the side thereof opposite the interior surface and surrounding and defining the passageway, and in which case the ceramic dielectric coating also covers the interior surface of the peripheral wall.

Survey of Patents (Corona Effect)
Organic Synthesis

Patent Number	Date of Issue	Inventor	Assignee	Title	Abstract
4,016,060	4/5/77	Frank E. Lowther Severna Park, MD	Union Carbide Corp. New York, NY	Corona Reaction Method and Apparatus	Corona-induced chemical reactions are conducted in a corona discharge zone in a corona discharge zone in which narrow high-voltage pulses are applied along with a relatively low-voltage bias potential. It is found that for many corona discharge reactions, such as the conversion of oxygen to the ozone, the present method increases the electrical efficiency of the reaction.
4,016,448	4/5/77	William L. Nighan Manchester, CT Water J. Wiegand Glastonbury, CT	United Technologies Corp., Hartford, CT	High-Power Electric Discharge Method and Apparatus for Laser Chemical Synthesis	Electric discharge method and appartus for producing a relatively large volume. It diffuses glow gas discharge plasma for laser, chemical synthesis and other applications at high-power, high-pressure and high-volume levels. Plasma properties are dominated by collision processes and the operational parameters of electric discharge are established to provide an average gas residence time, T_R, in a continuous wave convective electric discharge and a discharge duration time, T_p, in a pulsed electric discharge that is approximately equal to the plasma instability growth time T_G=CP/JE. P is the pressure

					in Torr, JE is the electric power desnsity in w/cm^3 and C is a constant typically about 10^{-4} for self-sustained gas-electric discharges and 10^{-3} for externally sustained gas-electric discharges. The convective electric discharge apparatus has a flow baffle system with a transverse turbulence generator plate for establishing an approximately isotopic and uniform turbulent flow pattern in the electric discharge region, so that T_R and T_G are representative of local values.
3,954,585	5/4/77	Richard J. Lagow Manchester, MA Lois L. Gerchman Stoneham, MA Robert A. Jacob Grand Forks, ND	Massachusetts Institute of Technology Cambridge, MA	Synthesis of Trifluoromethyl-Substituted Compounds	Trifluoromethyl-substituted compounds are formed in a corona discharge or glow discharge plasma of trifluoromethyl radicals from an organic trifluormethyl source. A substrate possessing easily replaceable ligands, such as halogen or carbonyl, is initially contacted either in the plasma and within a short distance from a downstream visible edge of the plasma or outside of the visible portion of the plasma and within a short distance from the downstream visible edge, to effect a substitution of the halogen or carbonyl ligand on the substrate with a trifluoromethyl radical without substantial decomposition of the substrate.

3,308,050	3/7/67	Michel Denis St. Germain-en-Laye France	Institut de Recherches de la Siderurgie Francaise, St. Germain-en-Laye France	Electric Discharge Apparatus for Chemically Reacting Flowing Gases	This invention relates to chemical reactors. It more particularly relates to apparatus for reacting gaseous materials. This invention provides apparatus for carrying out continuous reactions in a gaseous medium at very low temperatures and at low pressures with good yields in a single stage.
3,356,602	12/5/67	John A. Coffman Ballston Spa, NY William R. Browne Scotia, NY	General Electric Co., New York, NY	Process of Dimerizing Carboxylic Acids in a Corona Discharge	The invention relates to a process of treating carboxylic acids and esters thereof within a corona discharge.
3,494,845	2/10/70	Stephen W. Osborn Yardley, PA Edward Broderick, Perkasie, PA	Thiokol Chemical Corp., Bristol, PA	Method of Preparing Saturated Polymers of Simple Halocarbons in a Corona Discharge	Polymeric products containing a substantial amount of saturated components are obtained from tetrafluoroethylene, hexafluoropropylene, chlorodifluoromethane, and mixture thereof by subjecting these compounds to a corona discharge supported by electrodes spaced to define a gap of no more than 5 mm at a power density of 0.01-5 w/cm^2 of the average surface area of both of the electrodes.
3,475,308	10/28/69	James C. Burleson, St. Charles, MO William F. Yates, Chesterfield, MO	Monsanto Co. St. Louis, MO	Preparation of Alkylene Oxides in a Silent Electric Discharge	Alkylene oxides such as ethylene oxide and propylene oxide are prepared by subjecting gaseous mixtures of an olfin having a carbon chain with a corresponding number of carbon atoms and carbon dioxide under select conditions to a silent discharge.

2,689,217	9/14/54	William J. Cotton Butler, PA	Mathieson Chemical Corp., Virginia	Production of Chlorine Oxides by Electric Discharge	This invention relates to the production of valuable oxides of chlorine higher than the monoxide by direct oxidation of chlorine under the influence of electric discharge conditions. In a special aspect, it relates to the production of mixed oxides of chlorine predominating in chlorine dioxide. In another aspect it relates to the production of mixed oxides predominating in chlorine trioxide.
3,497,436	2/24/70	James C. Burleson, St. Charles, MO William F. Yates, Chesterfield, MO	Monsanto Co. St. Louis, MO	Manufacture of Alcohols Using a Silent Electric Discharge	Paraffin hydrocarbons are converted to alcohols by subjecting the paraffin hydrocarbons and carbon dioxide under select conditions to a silent electric discharge.
3,519,557	7/7/70	Carl E. Pruiss Madison, CT	Sun Oil Co. Philadelphia, PA	Controlled Hydrogenation Process	An improved process of hydrogenating petroleum fractions in which the temperature and reaction are controlled by intermittently dissipating the exothermic heat of reaction before the temperature rises above a predetermined level by carrying out the hydrogenation in several and separate catalytic reaction stages and interposing a slight, but rapid pressure reduction of about 50 psi. In this way, gaseous and very light materials such as H_2S, CH_4 and CO_2 are vaporized and removed.

3,513,044	5/19/70	Wataru Ishibashi Nosu, Japan	Iwatani & Co., Ltd. Osaka, Japan	Method for the Production of a High-Energy Fuel Employing Electric Spark Discharge	A method for the production of jet and rocket engine fuels, which comprises the steps of comminuting metals, alloys thereof or carbon into particles or pellets preferably of uniform sizes, and simultaneously disperses them in liquid hydrocarbon into a mixture. Electric spark discharges are induced by pairs of electrodes in the mixture among the pellets, whereby the pellets are cracked into gases and carbon particles until dispersions occur.
3,533,932	10/13/70	John A. Coffman Ballston Spa, NY William R. Browne, Scotia, NY	General Electric Co., New York, NY	Process of Dimerizing Carboxylic Acids in A Corona Discharge	The invention relates to a process of treating with a corona discharge a carboxylic acid having 0-3 ethylenic linkages and from 2-30 carbon atoms or esters of such an acid and an alcohol having 1-6 carbon atoms in a reducing atmosphere, preferably hydrogen, at a pressure of at least 1 atm. Mono- and triesters are exemplified.
3,207,582	9/21/65	Kiyoshi Inoue 182 3-chome Tamagawayoga-machi Setagaya-ku Tokyo Japan		Method of Synthesizing Diamond Particles by Utilizing Electric Discharge	The present invention relates to a method of synthesizing diamond particles from amorphous carbon or graphite and, more particularly, to a method of directly synthesizing diamond particles from amorphous carbon or graphite by utilizing high tempeature and impulsive high pressure produced by an electric discharge.

Survey of Patents
Surface Modification (Photography, Xerography, Printing)

Patent Number	Date of Issue	Inventor	Assignee	Title	Abstract
3,860,427	1/14/75	Shunji Matsuo Tokyo Shui Sato Tokyo Hidehiko Ishikawa Shinichi Nakamura Masahiko Taguchi	Konishiroku Photo Industry Co., Ltd. Tokyo, Japan	Quick-Processable Light-Sensitive Color Photographic Material	A light-sensitive silver halide photographic material comprising a polyethylene film base that has been subjected to a corona discharge treatment having coated theron a coupler-containing gelatin emulsion layer having a coupler-to-gelatin ratio in parts by weight of from about 1:3-1:5.
4,036,136	7/19/77	Tohru Takagi Hiratsuka, Japan	Kansai Paint Co., Japan	Corona Producing a Planagraphic Printing Sheet	A method for producing a plate or sheet useful in planographic printing, is characterized in that the surface of an oleophilic body comprising oleophilic resin is subjected to a corona discharge treatment to form a hydrophilic layer on said surface, and that said method can be carried out without difficulty at a low cost. The reproducibility of the printing plate thereby produced is excellent.
4,055,685	10/25/77	Gunther Bayer Hans Georg Fitzky Herbert Hoyer Gunter Luttgens Wolfgang Muller-Bardorff	Agfa-Gevaert Aktiengesellschaft West Germany	Process for the Improvement of the Adhesion of Photographic Layers to a Film Web by Means of a Corona Treatment	Corona treatment involves the continuous subjection of the film web to a steady alternating current discharge. This treatment is carried out in two consecutive stages, the corona discharge frequency, f_1 in the first stage being in the range f_1=5 to 100 kHz, and the corona discharge frequency, f_2, in the second stage being in the range f_2=500 to 200 kHz.

4,027,049	5/31/77	Yougoro Masuda Takashi Sato Shigeru Kawase Japan	Kanebo, Ltd. Tokyo, Japan	Process for Electrostatic Direct Transfer Printing of Designs on Fabrics	A process of electrostatically printing a design on fabrics. It comprises forming an electrostatic latent image corresponding to the design on a photoconductive material or an insulating dielectric material; developing the latent image with a toner comprising at least one chargeable resin and at least one colorant; transferring the developed toner image directly to a fabric having a surface inherent resistivity, as measured at a temperature of 20° C and a relative humidity of 65%, of about 1×10^6 to about 1×10^{14} ohms; and fixing the toner image to the fabric. When the transferring of the toner image is carried out using a corona discharge, a smooth sheet having a volume inherent resistivity of about 1×10^6 to about 1×10^{20} ohms and a dielectric constant of about 2 to about 40 can be overlaid on the fabric to apply corona discharge through the sheet and thereby produce better results.
3,994,725	11/30/76	Stewart William Volkers, Williamson, NY	Xerox Corp. Stamford, CT	Method for Enhancing Removal of Background Toner Particles	In an electrostatic copying apparatus, background toner particle removal is enhanced by employing a corona emission device between a development station and a background toner removal station. The corona emissions substantially reduce the background residual potential

					on a photoconductor and ensure that all background particles on the photoconductor are of a polarity opposite to the polarity of the charge on the latent image, whereby toner particles may be more easily removed from the photoconductor surface.
3,715,640	2/6/73	Masamichi Sato Asaka, Japan	Fuji Photo Film Co., Ltd., Kanagawa, Japan	Corona Charging Process and Apparatus in Electrophotography	The method of corona charging comprising a main corona discharging electrode positioned to act on an insulating layer on an electroconductive supporter which is not earthed, and corona discharge electrodes of a polarity opposite to said main electrode positioned to simultaneously act with said main electrode on a portion of the electroconductive supporter. In the structure of photographic material comprising the conductive layer on highly insulated supporter, it is difficult to earth the conductive layer.
3,760,152	9/18/73	Takashi Saito Tokyo, Japan Yujiro Ando Yokohama, Japan	Canon Kabushiki Kaisha Tokyo, Japan	Corona Discharge Device for Removing Dielectric liquid	A device for removing dielectric liquid from a holder member for a photosensitive medium or the like used in an electrophotographic copying machine. It includes a corona discharging electrode, a shield member substantially surrounding the corona discharging electrode in electricity isolated relationship with it and having at least a discharge opening, corona discharge converging means provided adjacent to the discharge opening of the shield member, and a power

					source from which a voltage is applied to the corona discharge electrode.
3,616,395	10/26/71	Leonard M. Carreira Penfield, NY Vsevolod Tulagia Rochester, NY	Xerox Corp. Rochester, NY	Photoelectrophoretic Imaging with Corona Field Application	A photoelectrophoretic imaging system and apparatus are described in which electically photosensitive particles in which dispersed in a carrier liquid are subjected to an electric field and exposed to imagewise light causing selective particle migration in image configuration. An electrostatic charge formed on the surface of the imaging suspension provides the electric field.
3,377,262	4/9/68	Michael Karickhoff Jerry Claud Lofland Circleville, OH	E. I. duPont de Nemours & Co., Wilmington, DE	Process of Improving the Printability Drafting Film Using an Electric Discharge and Heat	This invention relates to posttreatment of filled, synthetic, organic polymer shaped articles and, more particularly, to drafting films and processes for their treatment to affect the definition of applied ink markings on the surface.
3,418,115	12/24/68	Richard Menold Gerhard Heyl Water Shum West Germany	AGFA-Gevaert Aktiengesellschaft Leverkasen, West Germany	Bleaching out Electrophotographic Sensitizers by Simultaneous Exposure to Light and Corona Discharge	Dye-sensitized electrophotographic layer has the dye bleached out after image is developed by exposing the layer to dye-absorbing light in presence of corona discharge.
3,564,239	2/16/71	Teizo Kushima Japan	Minolta Camera Kabushiki Kaisha, Osaka, Japan	Flow-Type Photoelectric Duplicating Machine Having Means for Changing the Corona Voltage in Accordance with the Paper Speed.	In a flow-type photoelectric duplicating machine in which the original to be duplicated and the photosensitive paper are positioned in such a way as to be optically interrelated and connected by the optically connecting optical system. For this purpose the original that is illuminated by the illuminating device is gradually sent from one end to the other

across the light path of the said optical system, while synchronously the photosensitive paper is sent, after having been gradually sent from one end to the other through the electrostatic charging device, across the light path of the said optical system, so that the image of the original is gradually duplicated from one end to the other on the photosensitive paper; in such a flow-type photoelectric duplicating machine, an electric and/or mechanical connection is established between the passage speed of the photosensitive paper and the terminal voltage of the electrostatic charging device so they increase or decrease simultaneously.

Survey of Patents
Surface Modification (Semiconductors)

Patent Number	Date of Issue	Inventor	Assignee	Title	Abstract
3,806,365	4/23/74	Adir Jacob W. Roxbury, MA	LFE Corporation Waltham, MA	Process for Use in the Manufacture of Semiconductive Devices	A process step for use in the manufacture of semiconductor devices. To enable removal of all photoresist contamination, after development and etching of preselected portions of an oxide layer on a semiconductor slice, the material is exposed to a low pressure (few torr) rf generated "cold" plasma (200-300°C.), where the plasma is a homogeneous gaseous mixture of oxygen and organohalides. The organohalide preferably is a binary or ternary mixture where each component preferably includes no more than two carbon atoms per molecule and is desirably fully halogen-substituted. One of the substituents should include predominantly chlorine, while the other should include a predominance of either fluorine or fluorine-bromine combinations.
3,930,913	1/6/76	Adir Jacob Framingham, MA	LFE Corporation Waltham, MA	Process for Manufacturing Integrated Circuits and Metallic Mesh Screens	A process step for use in the manufacture of thin film integrated circuits, hybrid circuits and fine metallic mesh screens to enable removal of all organics and photoresist material from underlying

					metallic films without concomitant degradation of the metallic surface. After etching of preselected portions of an underlying critical metallic surface, the material is exposed to a low pressure (few torr) rf generated "cold" plasma, where the plasma is a homogeneous gaseous mixture of oxygen and nitrogen.
3,451,871	6/24/69	Heinrich Bauer Hans Romgens Monchen-Gladbach Hans-Jurgen Sontag Franz Zeppelzauer West Germany	Dessau Vetaphone Gesellschaft, Hambur, West Germany	Method of Treating Metallic Surfaces	A method of treating at least one surface of a metal member, such as an elongated sheet metal member, to improve the adherence of a cover layer in which the surface is subjected to the action of a high-voltage, high-frequency alternating field established between the member and a rotating electrode spaced from it.
3,951,843	4/20/76	Adir Jacob West Roxbury, MA	LFE Corporation Waltham, MA	Fluorcarbon Composition for Use in Plasma Removal of Photoresist Material from Semiconductor Devices	A organohalide composition for use in the manufacture of semiconductor devices. To enable the removal of all the photoresist material along with its inorganic contamination, after development and etching of preselected portions of an oxide layer on a semiconductor slice, the material is exposed to a low pressure (few torr) rf generated "cold" plasma (200-300°C), where the plasma is a homogeneous gaseous mixture of oxygen and organohalides. The organohalide preferably is a binary and ternary mixture where each component includes

					no more than two carbon atoms per molecule and is desirably fully halogen-substituted. One of the substituents should predominantly chlorine, while the other should include a predominance of either fluorine or fluorine-bromine combinations.
3,795,557	3/5/74	Adir Jacob West Roxbury, MA	LFE Corporation Waltham, MA	Process and Material for Manufacturing Semiconductor Devices	A process step and material for use in the manufacture of semiconductor devices. To facilitate the etching of unmasked silicon dioxide, silicon nitride, silicon monoxide, bare silicon layers, or various refractory metals on preselected portions of a semiconductor slice, the material is exposed to a low-pressure rf generated "cold" plasm (under 325° C) produced from a homogenous gaseous binary mixture of oxygen and a halocarbon. The halocarbon is preferably a gas having one carbon atom per molecule and is preferably fully fluorine-substituted.
3,951,709	4/20/76	Adir Jacob W. Roxbury, MA	LFE Corporation Waltham, MA	Process and Material for Semiconductor Photomask Fabrication	A process step and material for use in the manufacture of semiconductor photomasks. To facilitate the etching of unmask chromium, gold and other metals capable of forming oxychloride derivatives on preselected portions of a substrate material, the material is exposed to a low-pressure rf-generated "cold" plasma (under 300° C) produced from a homogeneous gaseous mixture of oxygen and a halogen-containing compound.

Survey of Patents
Surface Modification (Textiles, Fibers, Polymers)

Patent Number	Date of Issue	Inventor	Assignee	Title	Abstract
3,817,701	6/18/74	Walter J. Thorsen El Cerrito, CA	U.S. Secretary of Agriculture	Corona Treatment of Textiles	A process for enhancing the spinnability of fibers, particularly wool and cotton, which comprises exposing said fibers to a corona discharge zone for imparting both a transitory and permanent increase in cohesiveness to said fibers.
4,002,907	1/11/77	Klaus Kalwar Amshausen, Germany		Arrangement for Treating Sheet-Material Webs by Corona Discharge	An arrangement is provided for advancing a sheet-material web in a path, and a device for producing corona discharges includes at least one electrode roller and at least one cooperating counter roller, between which the web is compelled to pass. An edge position control controls the position of the edges of the web and includes portions of the electrode roller, which serve as guides for the edges of the web.
3,962,386	6/8/76	Gary L. Driscoll Boothwyn, PA	Sun Research and Development Co., Philadelphia, PA	Corona Discharge Treatment of Foam Fibrillated Webs	Nonwoven fabrics are produced from polyolefin foam fibrillated webs. The webs are treated with corona discharge to improve the adhesion when they are laminated under heat and pressure to form a nonwoven fabric.

4,028,551	6/7/77	Kenneth P. Thompson Canton, NC	Champion International Corp., Steamford, CT	Apparatus and Method for Corona Discharge Priming a Dielectric Web	A dielectric web substrate is primed for adhesion of a plastic coating by passing between two rolls in rolling contact therewith. One roll has a conductive core and an insulating jacket while the other roll has a conductive surface. A high-voltage ac source is connected between the conductive portions of both rolls to establish a corona discharge between to effect said priming. Additional rolls with conductive surfaces may be connected in parallel with the first such roll.
4,029,876	6/14/77	Theodore R. Beatty Harry Vourlis Ohio	Union Carbide Corp., New York, NY	Heat-Treated, Corona-Treated Polymer Bodies and a Process for Producing Them	Polymer films, such as polypropylene and polyethylene, having improved surface adhesion characteristics and produced by subjecting conventional polymer films to a heat treatment followed by a corona treatment at an energy density-to-film surface of at least 9 w-min/ft^2.
3,914,521	10/21/75	Theodore R. Beatty Harry Vourlis Ohio	Union Carbide Corp., New York, NY	Heat-Treated, Corona-Treated Polymer Bodies and a Process for Producing Them	Polymer films; such as polypropylene and polyethylene, having improved surface adhesion characteristics and produced by subjecting conventional polymer films to a heat treatment followed by a corona treatment at an energy density-to-film surface of at least 9 w-min/ft^2.

3,017,339	1/16/62	Bradley Dewey Cambridge, MA	American Can Co., New York, NY	Method of and Apparatus for Treating Surfaces of Extruded Thermoplastic Tubing with Corona Discharge	The present invention relates to the surface treatment of thermoplastic tubing for the reception of protective and decorative coating, printing inks, adhesives and the like in a manner which causes them to readily and firmly adhere, and has particular reference to a method of and apparatus for improving the adherence characteristics of exterior and interior surfaces of polyethylene or similar thermoplastic tubing by the application of a high-voltage electrical stress accompanied by corona discharges, as the tubing is extruded in substantially continuous lengths.
3,133,193	5/12/64	John E. Guillotte Thomas F. McLaughlin, Jr. Wilmington, DE	E. I. du Pont de Nemours & Co., Wilmington. DE	Corona Discharge Apparatus for the Surface Treatment of Plastic Resins	This invention relates to a novel apparatus for treating the surface of plastic articles with a corona discharge, and more particularly to an apparatus for the treatment of the surface of an article made from a hydrocarbon resin with a corona discharge to render the surface more printable. The apparatus is located on one side only of the plastic article.
4,059,497	11/22/77	Andreas Kolbe Peter Dinter West Germany	Hoechst Aktiengesellschaft West Germany	Corona Apparatus	A corona apparatus comprising at least roller electrode assembly, a supporting surface acting as a counter electrode, and a generator for generating an alternating current voltage for said electrode

					assembly characterized in that the roller electrode assembly comprises: a central shaft, an electrode, means for mounting said electrode to said shaft, a tube, means for mounting said tube for rotation about said shaft, said electrode contained within said tube, means for electrically connecting said electrode to said generator and means for supporting said shaft whereby said tube is positioned adjacent to said counter electrode.
3,661,735	5/9/72	Arthur H. Drelich Plainfield, NJ	Johnson & Johnson	Shaped Articles Having Improved Surface Properties and Corona Discharge Methods and Apparatus for Making the Same	Films, sheets, expanded or foamed products, fibrous webs, and similar synthetic, organic, polymeric shaped articles and structure having surfaces with improved bonding and adhesion capabilities increased fluid absorbency rates, and enhanced fluid wicking properties; and processes and apparatus for obtaining such improved capabilities and enhanced surface properties and characteristics by exposing said shaped articles to the action of an electrical corona discharge in an atmosphere of a reactive gas which does not have an affinity for electrons and which can also act as a chemical reducing agent, such as carbon monoxide.
3,222,188	12/7/65	Peter B. Feldman England	Monsanto Chemicals Ltd., London, England	Food Container Made of Synthetic Resin and Having Its Inner Surface Modified by Treatment with an	The invention relates to containers, especially for foods such as jam, jelly and syrup. Included in the invention is a pro- in which the inside surface of a container

				Electric Discharge	made from a thermoplastic styrene polymer is subjected to an electric discharge, so that when the container is filled with jam, jelly, syrup or other hot-filled edible substance, any tendency for bubbles to appear on the inside surface is substantially reduced.
3,546,065	12/8/70	Kurt W. Ostermeier Harrison, WI	Kimberly-Clark Corp., Neenah, WI	Corona Discharge Applied Directly to Creping Surface to Increase Adhesion of Web to Surface	A method and apparatus for the production of creped tissue paper. The normal adhesion of paper which dried on the surface of a crepe wadding dryer is increased, and controllably so, by applying a corona discharge to the dryer surface in advance of the application of the wet web. As is customary in such operations, the dryer is most suitably a rotating cylinder and the wet web is fed continuously to and creped from the cylinder. The adhesion of the web may be increased or decreased in the course of a creping operation to control the crepe structure of the paper product. The extent of the effect of the corona treatment on adhesion is related to the degree of adhesion that would normally occur in the drying of the sheet on the dryer surface.

3,549,406	12/22/70	William J. Ambusk Rochester, NY	Eastman Kodak Co. Rochester, NY	Process of Coating Polymer Surfaces Activated by Corona Discharge	Polymeric photographic supports, particularly polyethylene coated paper containing antistatic material, are coated with photographic emulsion layers by corona activation of the polymer surface using a current for generating the corona having a frequency greater than about 200 cps. The frequency of the power source is adjusted to the speed of the corona activation of the surface and the speed of the coating the surface to obtain a uniform coating. In a representative process, a 400-cycle corona is applied to the polymer surface at 125 fpm, and the emulsion layer coated thereon at the same speed.
3,729,672	1/14/71	Louis A. Rosenthal Donald A. Davis New Jersey	Union Carbide Corp. New York, NY	Apparatus for Film Treating	Apparatus is disclosed for the surface treatment of a plastic body by exposure to a high-intensity voltage accompanied by corona discharge employing a solid-state sonic frequency pulse generator.
3,880.966	5/29/75	Daniel Zimmerman Stephen Schulze Stanley Wolfowitz New Jersey	Celanese Corp. New York, NY	Corona-Treated Microporous Film	Crystalline elastic polymeric films having a crystallinity of at least 20% and an elastic recovery of at least 40% after subjection to a 50% extension at 25°C and 65% relative humidity are subjected to a corona discharge treatment rendering the films permeable. The films are subsequently rendered microporous by stretching and heat setting operations

					to further increase permeability. This treatment to render the crystalline elastic film microporous may be first used on the film before the corona treatment. The films are useful as filters, flow control devices, etc.
3,736,492	1/14/71	Louis A. Rosenthal Donald A. Davis New Jersey	Union Carbide Corp. New York, NY	Film-Treating Method	A method is disclosed for the surface treatment of a plastic body by exposure to a high-intensity voltage accompanied by corona discharge wherein said voltage is a sequence of alternating-directional, sonic frequency pulses of electrical voltage.
3,736,493	1/14/71	Louis A. Rosenthal Donald A. Davis New Jersey	Union Carbide Corp. New York, NY	Film-Treating Process	A process is disclosed for the surface treatment of a plastic body, employing an alternating voltage of high intensity accompanied by corona discharge, which comprises exposing said surface to high-frequency alternating voltage and accompanying corona discharge, the frequency of said alternating voltage being adjusted to effect surface treatment under optimum loading conditions at any selected constant voltage level.
3,736,494	7/12/71	Louis A. Rosenthal Donald A. Davis New Jersey	Union Carbide Corp. New York, NY	Appartus for Film Treatment	Method and appartus are disclosed for the surface treatment of a plastic body by exposure to a high-intensity voltage in the sonic frequency range accompanied by corona discharge, which comprises employing as said voltage a series

					of alternating-directional pulses of electrical voltage produced by energizing a pulse-forming network in which the requisite inductance and capacitance are provided exclusively by the transformer and the film-treating load.
3,703,569	11/21/72	Richard I. Wolkowicz Clarksboro, NJ	Scott Paper Co. Delaware Co., PA	Corona Treatment of Antistat Containing Sheet of Ethylene or Ethylene Copolymer	A method of preparing a sheet comprising thermoplastic polymeric material selected from polyethylene and copolymers of ethylene and propylene and an antistatic agent which includes the step of subjecting the sheet to corona treatment, whereby the static decay rate or the sheet is increased. The amount of antistat is at least the amount which gives an adequate static decay rate to the sheet after being modified by corona treatment but less than the amount sufficient to produce a measurable increase in the static decay rate of the sheet in the absence of corona treatment.
3,705,844	12/12/72	Gunter Haas West Germany	Kalle Aktiengesellschaft, West Germany	Apparatus for the Corona Treatment of Plastics	This invention relates to an apparatus for the corona treatment of shaped articles of plastic comprising a circuit including a generator for generating a voltage of appropriate frequency, a high-voltage transformer for producing a suitable operation voltage, a grounded electrode provided with a dielectric, at least one counter electrode and an inductance coil of variable inductivity.

3,754,117	8/21/73	Jacques Water Geneva, Switzerland	Agence Nationale de Valorisation de la Recherche Anvar, France	Device for Corona Treatment of a Layer of Plastic Material	A corona effect is produced across the layer of material to be treated by means of an intense electric field set up between two electrically insulated electrodes separated by a very narrow air-gap consisting of a slit formed at the extremity of an extruder head and through which the material is delivered continuously at its maximum temperature.
3,346,480	10/10/67	H. G. Schirmer Spartanbury, SC	W. R. Grace & Co. New York, NY	Apparatus for Corona Treatment of Polymer Film Including a Flexible Metal Foil Electrode	This invention relates to apparatus for the corona treatment of film.
3,405,052	10/8/68	H. G. Schirmer Spartanburg, SC	W. R. Grace & Co. New York, NY	Apparatus for Corona Treatment of Film Including a Porous Sintered Metal Electrode	This invention relates to appartus for the corona treatment of film.
3,391,044	7/2/68	Walter S. Kaghan William J. Schmitt Connecticut Percy M. Kay Ohio	Olin Mathieson Chemical Corp., New Haven, CT	Method for Improving Electric Glow Discharge Treatment of Plastic Materials	This invention relates to a method by which substantial improvement can be obtained in the electric glow discharge treatment of polyolefin, such as polyethylene or polypropylene, structures to improve the anchorage characteristic of a surface thereof.
3,485,734	12/23/69	James T. Pecka Tonawanda, NY	E. I. du Pont de Nemours & Co. Wilmington, DE	Process of Treating Polyimide Film in an Electric Discharge	Polyimide film containg fluoropolymer is treated by electrical discharge with an energy density of at least 300 w-min/ft^2 to impart adherability of a level to produce nonpeelable bonds.

3,222,188	12/7/65	Peter B. Feldmann England	Monsanto Chemicals Ltd., London, England	Food Container Made of Synthetic Resin and Having its Inner Surface Modified by Treatment with an Electric Discharge	The invention relates to containers especially for jam, jelly and syrup.
3,179,482	4/20/65	Paul Kassenbeck France	Institut Textile de France, Paris, France	Silent Electric Discharge Dyeing of Wool	The present invention relates to a process for the treatment by a liquid agent, particularly for dyeing and/or printing, of keratinuous fiber, especially wool.
3,661,735	5/9/72	Arthur H. Drelich Plainfield, NJ	Johnson & Johnson	Shaped Articles Having Improved Surface Properties and Corona Discharge Methods and Apparatus for Making the Same	Films, sheets, expanded or foamed products, fibrous webs and similar synthetic organic, polymeric shaped articles and structures having surfaces with improved bonding and adhesion capabilities, increased fluid absorbency rates, and enhanced fluid wicking properties; and processes and apparatus for obtaining such improved capabilities and enhanced surface properties and characteristics by exposing said shaped articles to the action of an electrical corona discharge in an atmosphere of a reactive gas which does not have an affinity for electrons and which can also act as a chemical reducing agent, such as carbon monoxide.
3,133,193	5/12/64	John E. Guillotte Thomas F. McLaughlin, Jr. Wilmington, DE	E. I. duPont de Nemours & Co., Wilmington, DE	Corona Discharge Apparatus for the Surface Treatment of Plastic Resins	This invention relates to a novel apparatus for treating the surface of plastic articles with a corona discharge, and, more particularly, to an apparatus for the treatment of the surface of an article made from a hydrocarbon resin with a

					corona discharge to render the surface more readily printable. The apparatus is located on one side only of the plastic article.
3,126,358	3/24/64	Jost Lemmerich West Germany	Siemens Schuckert-werke Aktiengesell-schaft, West Germany	Method for Improving the Breakdown Strength and Corona resistance of Olefin Polymers with Prehalogenized Hydro-carbons	This invention relates to olefin polymer products for dielectric or insulating pur-poses and, more particularly, to olefin polymers which, like polyethylene or polypropylene, have a median or average molecular weight of 30,000-130,000. However, in certain cases the invention is also applicable to high-density poly-ethylene up to molecular weights of 1,000,000.
3,075,040	1/22/63	Jost Lemmerich Ulrich Oestreich West Germany	Siemens-Schuckert-werke, Aktien-gesellschaft, West Germany	Method for Improving the Electric Strength and Flash-over or Glow-Discharge Resistance of Olefine Polymers	This invention relates to electric insula-ting materials of great dielectric strength and high flashover resistance for use at high voltages, such as in the order of kilovolts.
3,081,214	3/12/63	Thomas H. Strome Gambier, OH	Continental Can Co., New York, NY	Method of Bonding Poly-ethylene to Corona Discharge Treated Polyethylene Terephthalate	This invention relates to the bonding of polyethylene on substrate materials by prior treatment of the substrates before applying the polyethylene coating it.
3,113,208	12/3/63	George W. Traver Chicago, IL	E. I. du Pont de Nemours & Co. Wilmington, DE	Apparatus for and Method of Conditioning Plastic Surfaces for the Adhesion of Materials Coated Thereon	This invention relates to means for and method of treating the surfaces of bodies, comprising normally nonadherent material, to render them capable of ad-hering strongly to subsequently applied materials, such as printing ink, paint, pigment, cement, adhesives, floc and

					other materials, such as printing ink, paint, pigment, cement, adhesives, floc and other materials which it may be desired to coat, print, or otherwise attach on the treated surfaces, such subsequently applied materials, being called "coating materials."
3,294,971	12/27/66	Jack C. von der Heide Shelton, CT	National Distillers & Chemical Corp. New York, NY	Apparatus for the Corona Discharge Treatment of Thermoplastic Films Wherein the Discharge Electrode has a Plurality of Knife Edges	This invention relates to improvements in method and apparatus for treating plastic materials whereby the surface characteristics are modified to improve the capacity for the adherent reception of printing inks and to improve adhesion characteristics as between polyolefin extrusion coatings and various substrates.
3,546,065	12/8/70	Kurt W. Ostermeier Harrison, WI	Kimberly-Clark Corp. Neenah, WI	Corona Discharge Applied Directly to Creping Surface to Increase Adhesion of Web to Surface	A method and apparatus for the production to creped tissue paper. The normal adhesion of paper, which is dried on the surface of a crepe wadding dryer, is increased, and controllably so, by applying a corona discharge to the dryer surface in advance of the application of the wet web. As is customary in such operations, the dryer is most suitably a rotating cylinder and the wet web is fed continuously to and creped from the cylinder and the wet web is fed fed be increased or decreased in the course of a creping operation to control the crepe structure of the paper product. The extent of the effect of the corona

					treatment on adhesion is related to the degree of adhesion that would normally occur in the drying of the sheet on the dryer surface.
3,421,930	1/14/69	Jack R. Knox Wilmington, DE Karl-Heinz Teumer Chicago, IL	Continental Can Co. New York, NY	Condensation of Monomer and Low N-MER Vapors to Increase Polymerization Rates in a Corona Discharge	A polymeric coating is formed on a substrate by exposing the substrate to the atmosphere of a gaseous polymerizable monomer, the substrate being maintained at a temperature below the condensation temperature of the gaseous monomer and inducing the polymerization by a corona discharge operating at a pressure at least equal to atmospheric.
3,415,683	12/10/68	John A. Coffman, Ballston Spa, NY William R. Browne, Scotia, NY	General Electric Co. New York, NY	Corona Discharge Coating Processes	This invention relates to the discovery that organic materials in the vapor form may be deposited on substrates such as bare metal surfaces or paper-like materials in a corona reaction zone as adherent organic coatings resembling polymer films formed by conventional chemical techniques. These organic materials include vapors not generally polymerizable by conventional chemical techniques, such as methane, benzene and toluene, for example. Such substrates as the interior surface of metal cans or the outer surface of metal rodlike bodies or wires, kraft paper and asbestos-paper are specifically disclosed.

3,407,130	10/22/68	Robert B. Hailstone Wilmington, DE	E. I. du Pont de Nemours & Co. Wilmington, DE	Process of Treating Polyvinylbutyral Sheeting by an Electric Discharge in Nitrogen to Reduce Blocking	This invention relates to a process for preparing polyvinylbutyral sheeting, which has a decreased tendency to block. More specifically, this invention relates to treating polyvinylbutyral by electrical discharge in a nitrogen atmosphere and thereby obtaining a surface on the sheeting that has a decreased tendency to adhere to itself or to other similarly treated polyvinylbutyral sheets.
3,407,130	10/22/68	Robert B. Hailstone Wilmington, DE	E. I. du Pont de Nemours & Co. Wilmington, DE	Process of Treating Polyvinylbutyral Sheeting by an Electric Discharge to Reduce Blocking	This application is a continuation in part of application Ser. No. 296,655 filed July 22, 1963, now U.S. Patent No. 3,282,722 and application Ser. No. 521,600 filed Jan. 16, 1966. This invention relates to an improved polyvinyl butyral sheeting for use in laminated safety glass and, more specifically, to sheeting which, when it contains relatively low amounts of moisture, may be laminated to form safety glass having a high resistance to penetration.
3,503,859	3/31/70	Askolds A. Goncarovs Herbert O. Corbett John J. Quackenbush Connecticut	National Distillers & Chemical Corp. New York, NY	Corona Discharge Treatment of Polymer Film to Increase the Adhesion Characteristics Thereof	A plastic film or sheet-treating apparatus having a plurality of rollers through which a plastic web travels. Alternate rollers are movable from a plane on one side of the web to a position on the other side of the web. The web is threaded straight between the fixed rollers and the lowered movable rollers.

					The movable rollers are then moved upward to obtain the desired looped arrangement of the web with appropriate wrap of the web, web tension being obtained around the roller or rollers adjacent to one or more treating electrodes.
3,549,406	12/22/70	William J. Ambusk Rochester, NY	Eastman Kodak Co. Rochester, NY	Process of Coating Polymer Surfaces Activated by Corona Corona Discharge	Polymeric photographic supports, particularly polyethylene coated, containing antistatic material, are coated with photographic emulsion layers by corona activation of the polymer surface using a current for generating the corona having a frequency greater than about 200 cps. The frequency of the power source is adjusted to the speed of the corona activation of the surface and the speed of coating the surface to obtain a uniform coating. In a representative process, a 400-cycle corona is applied to the polymer surface at 125 fpm and the emulsion layer coated thereon at the same speed.
3,166,488	1/19/65	Alexander G. Makowski Fayville, MA	American Can Co. New York, NY	Apparatus for Treating Thermoplastic Containers by Corona Discharge	The present invention relates to the surface treatment of thermoplastic tubing containers for the reception of protective and decorative coating, inks, adhesives, etc. in a manner that causes them to readily and firmly adhere thereto, and has particular reference to an improved apparatus for treating the surfaces with an electrical corona discharge.

3,157,785	11/17/64	Frank E. Dobson Charles A. Badavas Richard S. Flint Massachusetts	American Can Co. New York, NY	Corona Treating of Hollow Plastic Articles	This invention relates to the surface treatment of hollow plastic articles to provide a surface receptive to inks, coatings, adhesives, etc. and especially to the interior treatment of hollow containers to ensure that interior coatings will adhere reliably and uniformly to them.
3,017,339	1/16/62	Bradley Dewey Cambridge, MA	American Can Co. New York, NY	Method of and Apparatus for Treating Surfaces of Extruded Thermoplastic Tubing with Corona Discharge	This relates to the surface treatment of thermoplastic tubing for the reception of protective and decorative coatings, printing inks, adhesives etc. in a manner that causes them to readily and firmly adhere thereto, and has particular references to a method of and apparatus for improving the adherence characteristics of exterior and interior surfaces of polyethylene or the like thermoplastic tubing by the application of a high voltage electrical stress accompanied by corona discharge as the tubing is extruded in substantially continuous lengths.

Survey of Patents
Surface Modification (Metals)

Patent Number	Date of Issue	Inventor	Assignee	Title	Abstract
3,452,565	7/1/69	Gilbert C. Cadwell Lakeside, CA	Rohr Corporation California	Electric Discharge Machine and Method of Metal Forming	Apparatus for forming metals by electric discharge in electroconductive water (tap water). Includes vacuum die and spark gap transducer structures that engage under pressure to clamp and seal a workpiece therebetween and respectively provide vacuum and water-filled cavities on opposite sides of the workpiece. The body of water in the transducer cavity provides the shock wave generating and transfer medium and empties by gravity when the formed part is removed from the die. The transducer cavity is filled from the bottom to thus force all entrapped air outwardly through a restricted water flow escape port at the top of the cavity. Shock waves for forming the workpiece are generated in the water medium by discharge of capacitively stored energy within the water and across the transducer discharge gap. The voltage level desired for charge and discharge of the bank is preset on a second voltmeter-type relay which follows the rise in voltage on the bank as the same charges, this relay having switch means which

					become actuated when the voltage the bank reaches the level preset on the relay. Acuation of the switch means at the desired voltage level causes sequential openings of the charging circuit to terminate the charge, disconnection of the bank from the charging circuit to isolate the power supply, the closing of a vacuum type discharge switch to transfer the charge to the discharge gap, and the closing of the grounding switch to dump any residual charge to ground. Sequentially operated time-delayed relays or, alternatively, a cyclincally operable multicam motor actuator are employed to effect the charge and discharge cycle of operations.
3,210,263	10/5/65	Wallace Jones Saltsburg, PA	Nuclear Materials & Equipment Corp. Apollo, PA	Electric Discharge Apparatus for Etching	This invention relates to electric discharge apparatus and has particular relationship to cathodic vacuum etchers such as are disclosed in application 786,516, now Patent Number 3,097,154 filed Jan. 13, 1959 to John E. Eck et al. for Apparatus and Method for Etching Objects and assigned to Nuclear Materials and Equipment Corporation.
3,190,772	6/22/65	Bernhard Berghaus Hans Bucek Switzerland		Method of Hardining Work in an Electric Glow Discharge	Hardening processes may be performed in an electric glow discharge of which the characteristic is that diffusion processes are accelerated in the ionized gas atmosphere, substances penetrating into

					the surface of the work from the gas owing to the ion bombardment. The articles to be treated are commonly connected so as to operate as electrodes and will therefore be covered by a glow while being heated to a predetermined temperature when the energy conversion is properly adjusted. A wide variety of hardening treatments may be performed since it is possible to select the composition of the gas atmosphere accordingly, by way of example, a nitriding hardening treatment in gases containing nitrogen, a boron or silicon treatment in gas atmospheres containing boron or silicon, and the like.
2,946,708	7/26/60	Bernhard Berghaus Hans Bucek Switzerland	Elektrophysikalische Anstalt Bernhard Berghaus, Vaduz, Liechtenstein	Nitriding with Electric Glow Discharge	This invention relates to the production of improved surface layers on tubes of iron, steel or iron alloys, more particularly gun tubes. Owing to the great stresses to which such tubes are exposed by corrosion, erosion, firing, etc., their surface is quickly attacked and destroyed. More particularly, gun tubes, owing to rapid repeated firing, are subject to considerable wear, so that they become useless after a short time and have to be replaced.
3,655,438	4/11/72	Henley F. Sterling Ware, England Richard C. Swann, N. Palm Beach, FL	International Standard Electric Corporation New York, NY	Method of Forming Silicon Oxide Coatings in an Electric Discharge	This is a method of depositing a coherent solid layer of an oxide of silicon deposited on a surface of a substrate by establishing a glow discharge adjacent to said

					surface in an atmosphere containing a gaseous compound of the element or elements comprising the material.
2,867,912	1/13/59	Mathew F. Kritchever Glencoe, IL	25% to Horace Dawson Evanston, IL	Method for the Decon-tamination or Metal Foils	This invention relates to the decontamination of metal foils and is particularly useful in the removal of light oils from foil film.
3,451,871	6/24/69	Heinrich Bauer Hans Romgens Hans-Jurgen Sontag Franz Zeppelzauer West Germany	Rheinische Blattmetall Aktien-gesellschaft West Germany	Method of Treating Metallic Surfaces	A method of treating at least one surface of a metal member, such as an elongated sheet metal member, to improve the adherence of a cover layer thereto, in which the surface is subjected to the action of a high-voltage, high-frequency alternating field established between the member and a rotating electrode spaced therefrom.
2,660,547	11/24/53	Robert G. Robertshaw London, England	General Electric Co. London, England	Method of Forming a Particle Coating on a Metal Component of an Electric Discharge Device	This invention relates to electric discharge devices. More particularly, the invention is concerned with methods of forming particle coatings on metal components of electric discharge devices.

Survey of Patents
Environmental Protection

Patent Number	Date of Issue	Inventor	Assignee	Title	Abstract
3,865,722	2/11/75	Patrick C. Stoddard Falls Church, VA		Corona Discharge Treatment of an Oil Slick	By subjecting an oil slick freely floating on a water surface to the action of a corona discharge, the oil tends to conglomerate and become cohesive. As a result, removal of the oil from the water surface is facilitated.
3,983,021	9/28/76	Jay M. Henis St. Louis, MO	Monsanto Co. St. Louis, MO	Nitrogen Oxide Decomposition Process	The removal of nitrogen oxides from gases by solids contact and electric discharge.
3,742,475	6/26/73	Leonard N. Liebermann La Jolla, CA Stanley H. Lai San Diego, CA	TIF Instruments. Inc. Miami, FL	Gaseous Impurity Detector Employing Corona Discharge Phenomenon	A method is disclosed of detecting gaseous impurities, particularly halogens, in an ambient atmosphere by repeatedly pulsing a pair of electrodes disposed in that atmosphere with a voltage sufficient to cause a corona discharge in the continuous corona region, and detecting the average (dc) current component of such discharge, changes which correspond to changes in the concentration of such gaseous impurities. Apparatus is disclosed for detecting such impurities in concentrations as low as 1 ppm.
3,967,131	6/29/76	Romuald E. Slipiec Northbrook, IL Ronald A. Schultz Berkeley, IL	Ozone, Inc. Deerfield, IL	Corona Discharge Ozone Generating Unit	An ozone-generating unit comprises a number of dielectric tube and electrode assembiles bundled together in confronting contacting relationship, each

dielectric tube and electrode assembly preferably comprising a number of concentrically related and spaced dielectric tubes. The ends of the dielectric tubes into which gas to be ozonized is to flow terminate at different points located progressively axially inwardly thereof proceeding from the outermost to the innermost dielectric tube. An outer cyclindrical conductive electrode is rigidly frictionally fitted over the outermost dielectric tubes. Supported on the inner surface of each dielectric tube is an inner cylindrical electrode made of a sheet of resilient flexible conductive mesh material initially deformed into a given curvature different from that of the cylindrical inner surface of the associated dielectric tube so that the inner cylindrical electrode expands into contact with the cylindrical inner suface of the associated dielectric tube. The dielectric tubes are spaced apart at each end of the dielectric tube assembly by circumferentially spaced pieces of flexible resilient plastic tubing sandwiched between the dielectric tubes.

Survey of Patents
Other Applications

Patent Number	Date of Issue	Inventor	Assignee	Title	Abstract
3,522,356	7/28/70	Leonard E. Olds Grand Island, NY Nickolas J. Themelis Quebec, Canada Fritz O. Wienert Niagara Falls, NY Murray C. Udy		Electric Furnace Corona Melting Process	The higher the power input to an electric furnace during smelting and/or melting operations, the higher the output will be. In conventional operation, slag foaming and refractory damage limit the power input, particularly in the later stages of a heat. In the present invention, the electrodes are exactly positioned for "corona" heating and, by correlating slag temperature, composition and viscosity, and charging rate, with the degree of metallization of the charge, power input and metal production are maximized. Foaming and refractory damage are avoided. The invention is particularly adapted to producing iron and steel from all types of iron and iron oxide-bearing charge materials.
3,938,345	2/17/76	Joji Yamaga Morio Jido Japan	Agency of Industrial Science & Technology, Tokyo, Japan	Cooling Method by Use of Corona Discharge	In cooling an article by the phenomenon of corona discharge caused when the article is placed opposite a high-potential electrode, a high electric potential is applied between the article and the electrode. This invention contemplates interposing a corona focusing ring between the

				article and the high-potential electrode. When high electric potential is also applied to the corona focusing ring, the avalanche of ions generated by the electrode in consequence of corona discharge is focused on the article by the corona focusing ring, with the result that local cooling is effected on the article.
2,683,689	7/13/54	Henry J. Nicholls Minneapolis, MN	Appartus for Treating Granular Material Through Action of Electrostatic Corona	This invention relates to apparatus and an improved method for treating granular and solid materials with ultraviolet rays and ozone produced through the phenomenon of electrostatic corona.
3,188,245	6/8/65	Kiyoshi Inoue Tokyo, Japan	Electric Discharge Heat Treatment of Metals in Electrolytes	This invention relates to improvements in the art of heating metal objects by creating an electric discharge at the surfaces of the object immersed in an electrolyte solution.
3,232,747	2/1/66	Kiyoshi Inoue Tokyo, Japan	Electric Discharge Heating of Metals in Electrolytes	This invention relates to improvements in the art of heating metal objects by creating an electric discharge at the surfaces of the object immersed in an electrolyte solution.
3,198,675	8/3/65	Kiyoshi Inoue Tokyo, Japan	Electric Discharge Heat Treatment of Metals in Electrolytes	This invention relates to improvements in the art of heating metal objects by creating an electric discharge at the surfaces of the object immersed in an electrolyte solution.

REFERENCES

1. Akutin, M. S., et al. *Plast. Massy* 6:2 (1960).
2. American Chemical Society, *Ozone Chemistry and Technology* (1959).
3. Bell, A. T., et al. *J. Appl. Pol. Sci.* 19:1911 (1975).
4. Bickerman, J. J. *J. Coll. Sci.* 6 (4): 349-359.
5. Blausten, B. "Chemical Reactions in Electrical Discharges," American Chemical Society, Washington, D.C. (1969).
6. Bradley, A., and J. P. Hammes. *J. Electrochem Soc.* 110:15 (1963).
7. Brown, P. F., and J. W. Swanson. *TAPPI* 54 (12):2013, (1971).
8. Brennan, M. E. *Chem. Commun.* 956 (1970).
9. Byrene, G.A., et al. *J. Soc. Dyers Col* 113 (1972).
10. Corey, E. J., and W. C. Taylor. *J. Am. Chem. Soc.* 86:3881 (1964).
11. Dittmer, G. "Plasma Polymerization of Coatings in Vacuum," *Ind. Res Develop.* 169-183 (September 1978).
12. Drahushuk, R. L., and H. M. Rosen. "Resource Recovery with Ozonation Systems," Union Carbide corp., Tonawanda, NY.
13. Dugdale, R. A., and J. P. Coad. "Sputter Ion Plating," on paper presented at the International Conference on Advances in Surface Coating Technology-London, 1976.
14. Evans, D. F. *Chem. Commun.* 367-368 (1969).
15. Fairless, C. M. H. "Studies on the Industrial Analytical Utility of a Direct Current Discharge Plasma," *Am. Lab.* 101-108 (March 1978).
16. Foner, S. N., and R. L. Hudson. "Metastable Oxygen Molecules Produced by Electrical Discharges," *J. Chem., Phys.* 25(3):601-602 (1956).
17. Franson, P. "Automating the Epitoaxial Process," *Electronics,* November 14, 1974).
18. Gallo. C. F. "Corona–A Brief Status Report," *IEEE Trans. Ind. Appl.* IA-13(6):550-557 (1977).
19. Gallo, C. F., and G. S. Castle. "Parametric Study of Ozone Generation by Coronas," *IEEE Trans. Ind. Appl.* IA-14(1):84-86 (1978).
20. Gellerstedt, F., et al. "Singlet Oxygen Oxidation of Lignin Structures," in *Singlet Oxygen Reactions with Organic Compounds and Polymers* B. Randy and J. F. Rabek, Eds. (New York:John Wiley & Sons, Ind., 1978).
21. Gillette, Hollahan and Carlson. *J. Vac. Sys. Technol.* 7:524 (1970).

22. Gleit, C. E. "Electronic Apparatus for Ashing Biologic Specimens," *Am. J. Med. Electronics* 2(2):112-118 (1963).
23. Gluskoter, H. J. "Electronic Low-temperature Ashing of Bituminous Coal," *Fuel* 44(4):285-287 (1965).
24. Goring, D. A. I. *Pulp Paper Mag (Canada)* 68:T-372, (1967).
25. Green, R. W. *TAPPI* 48(9):80A, (1965).
26. Hall, J. R., C. A. L. Westerdahl and M. J. Bodnar. "Activated Gas Plasma Surface Treatment of Polymers for Adhesive Bonding, Part II," Picatinny Arsenal Technical Report 4001, *J. Appl. Polymer Sci.* 13: 1085-2096 (1969).
27. Hauser, T. R., T. A. Hinners and J. L. Kent. "Atomic Absorption Determination of Cadmium and Lead in Whole Body by a Reagent Free Method," *Anal. Chem.* 44 (11):1819-1821, (1972).
28. Hines, R. A. "An Investigation of the Chemical Changes Occurring on the Surface of Polyethylene During Treatment to Render the Surface Printable," 132nd National Meeting ACS. New York, September 8-13, 1957, *J. Am. Chem. Soc: 17(2) (1958).*
29. Hollahan, J. R. "Analytical Applications of Electrodelessly Discharged Gases," in *Topics in Chemical Instrumentation,* S. Z. Lewin, Ed., New York University, New York.
30. Hollahan, J. R., and A. T. Bell. *Techniques and Applications of Plasma Chemistry* (New York: John Wiley & Sons, Inc., 1974).
31. Hollahan, J. R., and T. Weydeven. "Moisture Resistant and Anti-Reflection Optical Coatings Produced by Plasma Polymerization of Organic Compounds," *Proc. Soc. Optical Instrumentation Eng.,* Vol. 50 (1974).
32. Hollahan, J. R., et al. "Combination Moisture Resistant Antireflection Plasma Polymerized their Films for Optical Coatings," *Appl. Optics* 13 (August 1974).
33. Hudis, M., et al. "Plasma Polymerization of an Ethylene-Nitrogen Gas Mixture," *Polymer Lett* 13:549-557 (1957).
34. Irving, S. "A Dry Photoresist Removal Method," *Proc. Kodak Photoresist Seminar* (1968).
35. Kahn, A. U. *Science* 168: 476-577 (1970).
36. Kaplan, M. L. "Singlet Oxygen," *Chem. Technol.* 621-626, (1971).
37. Kaplan, M. L., et al. "Polydiene Oxidations with Singlet Oxygen," *Rubber Chem. Technol.* 423-434, (1972).
38. Kim, C. Y., et al. *J. Polymer Sci.* C 30: 533 (1970).
39. Kneip, T. J., M. Eisenbud, C. D. Strethlow and P. C. Freudenthal. "Airborne Particulates in New York City," *J. Air Poll. Control. Assoc.* 20(3):144-149 (1970).
40. Kometani, T. Y., J. L. Bove, B. Nathanson, S. Siebenberg, and M. Magyar. "Cry Ashing of Airborne Particulate Matter on Paper and Glass Fiber Filters for Trace Metal Analysis by Atomic Absorption Spectometry," *Environ. Sci. Technol.* 6(7):617-620 (1972).
41. Kronick, P. L., et al. *J. Polymer. Sci.* Al(7):767 (1969).

42. Lam, D. A., et al. "A Mechanism and Kinetics Study of Polymeric Thin-Film Deposition in Glow Discharge, *J. Macromol. Sci.-Chem.* A 10(3):421-450 (1976).
43. LFE Corporation. "Plasma Systems for High Yield Microcircuit Fabrication."
44. Lightner, D. A., et al. "The Dye Sensitized Photooxygenation of Kryptopyrrole," *Experientia* 29:262-264 (1973).
45. Lilienfeld, P. "Coherent Ion Plasma Oscillations in a DC Corona Discharge in Rare Field Air," *IEEE Trans, Ind. Appl* IA-13(4):374-378 (1977).
46. Lindsay, P. Private communication (1978).
47. List, G. R., R. L. Hoffmann, W. F. Kwolek and C. D. Evans "Determination of Copper in Edible Soybean Oils," *J. Am. Oil. Chem. Soc.* 45:872-875 (1968).
48. Matheson, I. B. C., and J. Lee. *Chem. Phys. Lett.* 7(4):475-476 (1970).
49. McKelvey, J. M. "Polymer Processing." in *Surface Phenomena.* (New York: John Wiley & Sonx. Inc., 1962).
50. McLaughlin, T. F. "Surface Characteristics of Polyolefin Films and their Influence on Printability," paper presented at 8th Annual meeting Flexographic Technical Association, 1966).
51. Merkel, P. B. and D. R. Kearns. *Chem. Phys. Lett.,* 12(1):120-122, (1971).
51a. Moore, A. O. *Electrostatics* (New York: Doubleday and Co., Inc., 1968).
52. Mulford, C. E. "Low Temperature Ashing for the Determination of Copper in Edible Soybean Oils," *J. Am. Oil. Chem. Soc.* 45:872-875 (1968).
53. Murray, J. S., and J. R. Orr, Eds. *Ozone Chemistry and Technology: A Review of the Literature, 1961-1974* (The Franklin Institute Press, 1975).
54. Murray, R. W., and M. L. Kaplan. *J. Am. Chem. Soc.* 91(19):5358-5364 (1969).
55. Murray, R. W., M. L. Kaplan, J. W. P. Lin, W. C., Lumma, R. D. Simelaha. *Organic Ozone Reactions as Singlet Oxygen Sources,* 2nd Internat. Clean Air Congress, CPIC.
56. Nasser, E. *Fundamentals of Gaseous Ionization and Plasma Electronics,* (New York: Wiley Interscience, 1971).
57. O'Kane, D. F., and K. L. Mittal. "Plasma Cleaning of Metal Surfaces," *J. Vac. Sci. Technol.* 11(3):567-569 (1974).
58. Pavlath, et al. *Appl Polymer Symp.,* Part II, 18:1371 (1971).
59. Poulsen, R. G. "Plasma Etching in Integrated Circuit Manufacture-a Review," Bell Northern Research, Ottawa, Canada, (1976).
60. Ranby, B., and J. F. Rabek. *Singlet Oxygen, Reactions with Organic Compounds and Polymers, Eds.* (New York: John Wiley & Sons, Inc., 1978).
61. Rosen, H. M. "Ozonation and the Safe Drinking Water Act, Plus," Union Carbide Corp., Tonawanda, NY.

62. Rosen, H. M. "Wastewater Ozonation: A Process Whose Time Has Come," *J. Am. Soc. Civil Eng.* 65-69 (March 1976).
63. Schaap. A. P. *Singlet Molecular Oxygen* (New York: Halsted Press, 1976).
64. Sheer, C. "Arc Jets in a Science and Technology," *Vistas in Science* (1968).
65. Seegar, M., et al. "Analysis of Plasma Polymerized Hydrocarbons by Pyrolysis/Gas Chromatography," *J. Polymer Sci.* 15:1403-1411 (1977).
66. Shuhet, J. L. *The Palsma State,* (New York: Academic Press, 1971).
67. Smith, L., et al. "Glow Discharge Surface Treatment for Improved Cellular Adhesion," paper presented at Polymer Chemistry Div., 170th ACS National Meeting, Chicago, IL August 27, 1975.
68. Smith, L. "Measurement of Cellular Adhesion on Glass and Polymer Substances," Master's Thesis, Div. of Materials Science, Dept. of Mechanical Engineering, University of Utah, (1973).
69. Smolinsky, G., et al. *Div. Org. Plastic. Chem. Preprints* 28(1):537 (1968).
70. Snelling, D. R. *Chem. Phys. Lett.* 2(5):346-348 (1968).
71. Stephenson, L. M., et al. *J. Am. Chem. Soc.* 95:3074 (1973).
72. Stillwagon, J. E., and L. M. Butcher. "Electrostatic Treated Full in Coatings," *Laminations,* Part 2, *Modern Converter* (1971).
73. Stone, R. B., and J. R. Barrett. *Jr. Textile Bull.* 88:65 (1962).
74. Tibbitt, J. M., et al. "A Model for Kinetics of Plasma Polymerization, Mabromolecules," 10:647 (1977).
75. Thoresen, W. J. *Textile Res. J.* 41:331 (1971).
76. Truscott, T. G., et al. "In Vitro Photochemistry of Biological Molecules," *Photochem. Photobiol.* 17:43-51 (1973).
77. Tsuji, O., et al. "Preservation of the Plasma Ash Pattern of Biological Specimens by Subsequent Coating with a Plasma Polymerized Thin Film," *Murochem. J.* 22:229-235 (1977).
78. Wasserman, H. H. In: *Singlet Oxygen,* R. W. Murry, ed. (New York: Academic Press, 1978).
79. Weydeven, T. "Plasma Polymerized Coating for Polycarbonate; Single Layer, Abrasion Resistant, and Antireflection," *Appl. Optics.* 16(3): (1977).
80. Weydeven, T., et al. "Antireflection Coating Prepared by Plasma Polymerization of Perfluorobutene-2," *Appl. Optics.* 15(1):(1976).
81. Wiles, D. M. "Stabilization of Polymers against Singlet Oxygen," in Ranby and Rabek, *Singlet Oxygen, Reactions with Organic Compounds and Polymers,* B. Randy and J. F. Rabek, Eds. (New York: John Wiley & Sons, Inc., 1978).
82. Wolters, E., Th. M., et al. "Improving Self-Adhesion of EPDM by Means of a Singlet Oxygen Reaction," in *Singlet Oxygen, Reactions with Organic Compounds and Polymers,* B. Randy and J. F. Rabek, Eds. (New York: John Wiley & Sons, Inc., 1978).

83. Yang, M. G., K. M. Koliwad and G. E. McGuire. "Auger Electron Spectroscopy of Cleanup-Related Contamination on Silicon Surface," *J. Elect. Soc.* 675-678 (May 1975).
84. Yasuda, H. "Plasma for Modification of Polymers," *J. Macromol. Sci. Chem.*, A10(3):383-420 (1976).
85. Yasuda, H., et al. "Some Aspects of Plasma Polymerization of Fluorine Containing Organic Compounds, *J. Polymer. Sci.*, Polymer Chemistry Edition, 15:2411-2425 (1977).
86. Yasuda, H., et al. "Some Aspects of Plasma Polymerization Investigated by Pulsed R. F. Discharge," *J. Polymer. Sci.* 15:81-97 (1977).
87. Yasuda, H., et al., *J. Polymer. Sci.* 15:2749-2771 (1977).
88. Yasuda, H., et al. "ESCA Study of Polymer Surface Treated by Plasma," *J. Polymer. Sci.* 15:991-1019 (1977).
89. Yasuda, H., et al. *J. Appl. Polymer. Sci.* 17:201 (1973).
90. Yasuda, H. Conference on Polymeric Materials for Unusual Service Conditions, Ames Research Center, NASA, Moffett Field, CA, November 29-December 1, 1972.
91. Zisman, W. A. "Relation of the Equilibrium Contact Angle to Liquid and Solid Constitution," in *Contact Angle, Wettability, and Adhesion, Advances in Chemistry Series No. 43* (Washington, D.C.: American Chemical Society, 1964).

INDEX